# ESSENTIAL MATH SKILLS
# FOR ENGINEERS

DATE DUE

# ESSENTIAL MATH SKILLS FOR ENGINEERS

**Clayton R. Paul**

*Professor of Electrical and Computer Engineering*
*Mercer University*
*Macon, Georgia*

*and*

*Emeritus Professor of Electrical Engineering*
*University of Kentucky*
*Lexington, Kentucky*

**Celebrating 125 Years**
of Engineering the Future

**WILEY**

A JOHN WILEY & SONS, INC., PUBLICATION

Published by John Wiley & Sons, Inc., Hoboken, New Jersey.
Published simultaneously in Canada.

For general information on our other products and services or for technical support, please contact our Customer Care Department within the United States at (800) 762-2974, outside the United States at (317) 572-3993 or fax (317) 572-4002.

Wiley also publishes its books in a variety of electronic formats. Some content that appears in print may not be available in electronic formats. For more information about Wiley products, visit our web site at www.wiley.com.

*Library of Congress Cataloging-in-Publication Data:*

Paul, Clayton R.
   Essential math skills for engineers / Clayton R. Paul.
      p.  cm.
   Includes bibliographical references and index.
   ISBN 978-0-470-40502-4
1. Mathematics.  2. Engineering mathematics.  I. Title.
   QA37.3.P39 2009
   620.001′51—dc22

                                        2008042568


10 9 8 7 6 5 4 3 2

*To the Humane and Compassionate Treatment of Animals*

# CONTENTS

# PREFACE

The purpose of this brief textbook is to enumerate and discuss *only those mathematical skills that engineers use most often.* These represent the math skills that engineering students and engineering practitioners must be able to recall immediately and understand thoroughly in order to become successful engineers. We will not stray into esoteric topics in mathematics. If a math topic is not encountered frequently in the daily study or practice of engineering, it is not discussed in this book. The goal of the book is to give the reader lasting and functional use of these essential mathematical skills and to categorize the skills into functional groups so as to promote their retention.

Students studying for a degree in one of the several engineering specialities (i.e., electrical, mechanical, civil, biomedical, environmental, aeronautical, etc.) are required to complete numerous courses on general mathematics. This material represents a rather formidable amount of mathematical detail and can leave a student overwhelmed by its sheer volume. However, in the student's everyday studies in engineering, the vast majority of these skills and concepts are never used or are used so infrequently that they are quickly forgotten. This also applies to the engineering professional. The majority of the math skills that a student will use on a frequent basis represents only a small portion of all the math topics studied in math courses. If the student is distracted in a particular engineering course by struggling to remember these frequently encountered math skills, learning the particular engineering topic being studied will not take place. This book is intended to cover only those math skills used most frequently by engineering undergraduate students and engineering professionals. These few but critically important skills must be firmly understood and easily recalled by both students and practitioners if they are to be successful in their engineering studies and their future engineering practice.

I am frequently asked the question: "What do engineers do?" I have reduced my answer to the following succinct summary:

Engineers develop and analyze mathematical models of physical systems for the purpose of designing those physical systems to perform a specific task.

The next question that is asked is: "Why do they do that?" My answer to this question is:

Engineers do that so that they can develop *insight* into the behavior of a physical system in order to construct a mathematical model which when implemented in a physical system will accomplish certain design goals.

Hence, *mathematics is at the heart of engineering design*. This makes it abundantly clear that any student or practitioner of engineering must be fluent in the mathematical skills that they encounter frequently. Being able to use mathematical skills alone will not make a person a competent engineer, but not being able to use those skills will handicap their ability to become a competent engineer.

Chapter 1 covers miscellaneous math skills, such as writing the equation of a straight line, various formulas for the area and volume of common shapes, obtaining the roots of a quadratic equation, logarithms, reducing fractions via lowest common denominators, long division, trigonometry, complex numbers and algebra, common derivatives and integrals, and numerical integration.

Chapter 2 discusses the solution of simultaneous, linear, algebraic equations. Cramer's rule, Gauss elimination, and basic matrix algebra are also described. Chapter 3 covers solution of the most common form of ordinary differential equations: the linear, constant-coefficient, ordinary differential equation. Although nonlinear and/or non-constant-coefficient differential equations are encountered in engineering, the linear, constant-coefficient differential equation is the type encountered most frequently.

In the last twenty years, discrete-time systems such as computers have become very common. These systems are described by difference equations. In Chapter 5 we discuss the most common type of difference equation: the linear, constant-coefficient, difference equation, and in Chapter 6 we discuss the solution of linear, constant-coefficient, partial differential equations.

In the remaining three chapters we discuss important specialized solution topics. The Fourier series and Fourier transform are discussed in Chapter 7. In Chapter 8 we describe the Laplace transform method of solving ordinary as well as partial differential equations. Finally, in Chapter 9 we discuss briefly vector algebra and elementary vector calculus.

This text will provide readers with a compact but complete summary of all the mathematical skills that engineering students and professionals need in order to become successful engineers. Without a thorough and immediate understanding of these skills, one cannot expect to become a successful engineer.

*Macon, Georgia* CLAYTON R. PAUL

# 1 What Do Engineers Do?

This textbook is concerned with the mathematical skills that are essential to students and practitioners of all branches of engineering. The primary purpose of this brief book is to enumerate and discuss *only those mathematical skills that engineers use most often.* These represent the math skills that both engineering students and engineering practitioners must be able to recall immediately and understand thoroughly to be successful engineers. We will not stray into esoteric topics in mathematics. If a topic is not encountered frequently in the daily study or practice of engineering, it is not discussed here. The goal of this book is to give the reader lasting and functional use of these essential mathematical skills, and to categorize the skills into functional groups to promote their retention.

Students studying for a batchelor of science degree in one of the several engineering specialties (i.e., electrical, mechanical, civil, biomedical, environmental, aeronautical, etc.) are required to complete several courses in general mathematics. Prior to enrolling in a university, students are generally required to have completed high school courses covering algebra and trigonometry. At the author's institution, which is fairly typical, students are required to complete a semester of differential calculus and a semester of integral calculus in the freshman year. In addition, a course in differential equations is required to be completed in the sophomore year. Many engineering programs also require a semester in the junior year on multivariable mathematics. All of this mathematical material represents a rather formidable amount of detail and can leave a student overwhelmed by its sheer volume. However, in a student's everyday studies in engineering, as well as in professional engineering practice, the vast majority of those skills and concepts are never used or are used so infrequently that they are quickly forgotten. The majority of the math skills that students will use on a frequent basis represents only a small portion of the math topics studied in their math courses. It therefore makes sense for students to concentrate on and commit to memory and immediate usability only those skills that are used frequently, deferring those math topics that will be used infrequently to being "looked-up" when they arise. The healthy view of the role of mathematics in engineering is as a tool to understand the behavior of the particular engineering system being studied—in the same way that language is a means of communicating. If the student is distracted in a particular

*Essential Math Skills for Engineers,* By Clayton R. Paul
Copyright © 2009 John Wiley & Sons, Inc.

engineering course by struggling to remember frequently encountered math skills, learning the particular engineering topic being studied will not take place. This book is intended to cover only those math skills used most frequently by engineering undergraduate students and the majority of engineering practitioners.

Even though students do not use all these math skills on a frequent basis, they nevertheless benefit from being exposed to the majority of the mathematical concepts and mathematical sophistication they will study in their math courses; engineering study is also intended to be an "education." Students who go on to graduate school and pursue advanced degrees in engineering will need more math skills and topics the farther they go. However, the success of students in a graduate engineering program will also rely primarily on their having a solid understanding of the undergraduate program, which requires a solid understanding of the basic mathematics covered in this book. But it is important to remember that the majority of undergraduate engineering students will not pursue a graduate engineering degree. Hence, both students and practitioners must firmly understand and be able easily to recall these few but critically important math skills if they are to be successful in their engineering studies and in their future engineering practice. Students tend to be "overwhelmed" by the vast body of mathematical knowledge they have encountered in their math courses and are not readily able to compartmentalize which skills are most important to commit to immediate memory. They have not completed their studies and are therefore not able to discern this for themselves. We teachers of engineering easily forget this important fact. We have been studying engineering for many years and it is abundantly clear which math skills are frequently encountered. Students do not possess our broad overall view of the situation.

It is for this reason that I have undertaken to write this brief textbook on the essential math skills required of engineers. I have been studying engineering for over 50 years and have been teaching the subject for over 40 years. Although the majority of that time was spent in academia, I have also spent a substantial portion of it involved in the concurrent practice of engineering in industry. It has become very clear to me after this lengthy experience that the primary attribute which determines whether a student will be successful in his or her engineering studies is the person's ability for immediate recall and successful use of this small but important body of math skills. This observation also applies to engineering professionals.

I am frequently asked the question: "What do engineers do?" I have reduced my answer to the following succinct summary:

---

Engineers develop and analyze mathematical models of physical systems for the purpose of designing those physical systems to perform a specific task.

So the unique function of engineers is *design*. The next question that is asked is: "Why do they do that?" My answer to this question is:

> Engineers do that so that they can develop *insight* into the behavior of the physical system in order to construct a mathematical model which when implemented in a physical system will accomplish certain design goals.

Hence, *mathematics is at the heart of engineering design*. This makes it abundantly clear that any student or practitioner of engineering must be fluent in the mathematical skills that they encounter frequently. Being able to use mathematical skills alone will not make you a competent engineer, but not being able to use those skills will handicap your ability to become a competent engineer.

Engineering systems and the design problems associated with them are much too complicated today (and will no doubt become increasingly so in the future) to be understood by one's ordinary "life experiences." In fact, if this were so, engineering companies would not be at the top of industry jobs paying the highest salaries. Many students today seem to believe that the digital computer will make engineering easy and obviate the need for them to learn mathematics. Numerous computer programs exist that can solve (give a numerical answer to) the complicated mathematical equations that describe engineering systems. If the design of engineering systems were that simple, industries could, perhaps, simply pay minimum wage to someone and train the person (in a very short time) to use that computer program. But that would be missing the point of what engineers do. These computer programs give a numerical "answer," but they give little, if any, insight into how a particular engineering system behaves. To obtain that insight (which is critical to design), we must understand what the mathematics governing that system is telling us about its behavior. Without that understanding we would have to construct a physical prototype of the anticipated system and change the structure and parameters of that prototype endlessly and randomly, with no clear indication of how to arrive at an optimal solution to the design problem desired.

It is vitally important to keep in mind the important fact that the physical laws that govern the behavior of all physical systems are given in mathematical terms. Therefore, if we are to understand how to design that physical system to perform a particular task, we must understand how to solve these governing mathematical equations.

In the remaining part of this chapter I give examples of engineering systems and the equations that govern them. Examples such as these pervade all branches of engineering and are too numerous to give here, so I concentrate on a few disciplines: electrical engineering, mechanical engineering, and civil engineering. I do not go into how the equations governing a particular system are derived from the specific laws of the discipline, as that is the goal of your particular engineering courses. The goal in this book is to give you the mathematical tools to solve those equations in order to be able to understand what they are telling you about the behavior of that system.

***Example 1 (Electrical Engineering)***    Electrical engineers are concerned with electrical systems whose important variables—voltage, current, and power— are to be determined for a particular interconnection of electric elements (an electric circuit). Electrical systems are governed by Maxwell's equations. These are complicated partial differential equations (Chapter 6). However, they can be approximated for a large majority of electrical systems by two simple governing equations: Kirchhoff's voltage law (KVL), which governs the voltages of a particular interconnection of electrical elements that comprise an electric circuit, and Kirchhoff's current law (KCL), which governs the currents of the circuit. The individual electrical elements of the circuit (resistors, capacitors, inductors, diodes, transistors, etc.) relate the element's voltage to its current.

The electric circuit shown in Figure 1.1 consists of a voltage source and resistors. It is desired to determine the value of resistor $R$ such that the output voltage is $V = 3V$. The circuit can be "solved" by writing the mesh-current equations as

$$7I_1 - 3I_2 = 10$$
$$-3I_1 + (5 + R)I_2 = 0$$

These are two simultaneous, linear, algebraic equations that can be solved using Cramer's rule (Chapter 3) as

$$I_2 = \frac{\begin{vmatrix} 7 & 10 \\ -3 & 0 \end{vmatrix}}{\begin{vmatrix} 7 & -3 \\ -3 & (5+R) \end{vmatrix}} = \frac{30}{26 + 7R}$$

The desired voltage is $V = 3V$ and $V = RI_2$. Substituting $I_2 = 3/R$ gives $R = 78/9\,\Omega$.

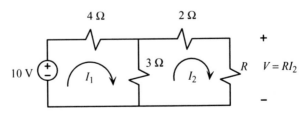

**Fig. 1.1.** Example 1: electrical engineering.

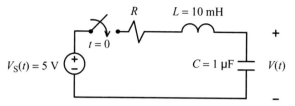

**Fig. 1.2.** Example 2: electrical engineering.

***Example 2 (Electrical Engineering)*** The electric circuit shown in Figure 1.2 is composed of a 5-V voltage source, a resistor whose value is to be determined, a 10-mH inductor, a 1-μF capacitor, and a switch that closes at $t = 0$. The design task here is to determine the value of $R$ and perhaps add other components so that the output voltage $V(t)$ rises smoothly and rapidly to 5 V at $t = 1$ ms. The differential equation relating the output voltage to the source voltage is

$$\frac{d^2V(t)}{dt^2} + \frac{R}{L}\frac{dV(t)}{dt} + \frac{1}{LC}V(t) = \frac{1}{LC}V_S(t)$$

The complete solution requires two initial conditions. The initial conditions at $t = 0^+$ (just after the switch closes) are $V(0^+) = 0$ and $dV(t)/dt|_{t=0^+} = 0$. This is a second-order, linear, constant-coefficient, ordinary differential equation whose solution we determine in Chapter 4. The general form of the solution is

$$V(t) = C_1 e^{p_1 t} + C_2 e^{p_2 t} + 5 \qquad t \geq 0$$

where the (as yet) undetermined constants $C_1$ and $C_2$ are determined by applying the two initial conditions. The exponents of the exponential terms, $p_1$ and $p_2$, are roots of the quadratic equation

$$p^2 + \frac{R}{L}p + \frac{1}{LC} = (p - p_1)(p - p_2) = 0$$

We solve this quadratic equation in Chapter 2, giving the two roots as

$$p_1, p_2 = -\frac{R}{2L} \pm \frac{1}{2}\sqrt{\left(\frac{R}{L}\right)^2 - \frac{4}{LC}}$$

In Chapter 2 we also investigate the exponential function $e^{pt}$, which arises quite frequently throughout engineering. There are three possibilities for these two roots:

Case I: roots real and distinct:

$$p_1 \neq p_2 \quad \text{if} \left(\frac{R}{L}\right)^2 > \frac{4}{LC}$$

Case II: roots real and equal:

$$p_1 = p_2 = p \quad \text{if} \left(\frac{R}{L}\right)^2 = \frac{4}{LC}$$

Case III: roots complex:

$$p_1, p_2 = \alpha \pm j\beta \quad \text{if} \left(\frac{R}{L}\right)^2 < \frac{4}{LC}$$

Cases I and II rise smoothly to the final value of 5 V, but case II does so very rapidly. Its general solution is

$$V(t) = C_1 e^{pt} + tC_2 e^{pt} + 5$$

However, case III has an oscillatory behavior, with the solution oscillating about the final value of 5 V but eventually settling down to the desired value of 5 V:

$$V(t) = K_1 e^{\alpha t} \sin \beta t + K_2 e^{\alpha t} \cos \beta t + 5$$

This is a very undesirable result, so we choose case II. For the two roots to be identical, we must have

$$\left(\frac{R}{L}\right)^2 = \frac{4}{LC}$$

or

$$R = 2\sqrt{\frac{L}{C}}$$

The two roots will be $p_1 = p_2 = p = -R/2L$ and hence are negative, so that the two exponential pieces of the solution will decay to zero, leaving the desired solution of 5 V. These two exponential parts of the solution will have decayed sufficiently to be considered close enough to zero in a time of approximately

$$\frac{5}{|p|} = 10\frac{L}{R} = 1 \times 10^{-3} \text{ s}$$

Substituting the value of $L = 10 \times 10^{-3}$ gives the desired value of $R = 100\,\Omega$. Solving $R = 2\sqrt{L/C}$ with the now known values of $R$ and $L$ requires that $C = 4\,\mu\text{F}$. Therefore, we must add another 3-$\mu$F capacitor in parallel with the existing 1-$\mu$F capacitor. (Capacitors in parallel add like resistors in series.) A plot of the response with these values from $t = 0$ to $t = 2$ ms is shown in Figure 1.3, indicating a successful design.

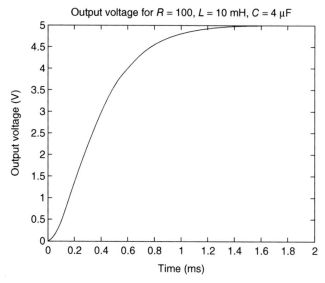

**Fig. 1.3.** The solution for repeated roots: $R = 100\,\Omega$, $L = 10\,\text{mH}$, $C = 4\,\mu\text{F}$.

The plots in Figure 1.4 compare how the solution behaves for

$R = 100\,\Omega$,   $L = 10\,\text{mH}$,   $C = 4\,\mu\text{F}$   (case II: the desired solution)

$R = 100\,\Omega$,   $L = 10\,\text{mH}$,   $C = 10\,\mu\text{F}$   (case I: distinct roots)

$R = 100\,\Omega$,   $L = 10\,\text{mH}$,   $C = 0.05\,\mu\text{F}$   (case III: the oscillatory solution)

This example has illustrated clearly that being able to solve the differential equation where the coefficients are symbols (one of the objectives of this book) as opposed to a numerical solution (solving the differential equation where the coefficients are numbers) has allowed us to go immediately to the optimum solution for the design rather than guessing a set of element values and solving for a numerical result, changing our guess, and re-solving and continuing in a never-ending fashion with no hope of arriving quickly at the optimum solution. This clearly illustrates that simply being able to solve a specific differential equation whose coefficients are numbers using a computer is not the answer to an effective and efficient method of engineering design. A numerical computer solution is simply an analysis of a given situation; design is the synthesis of a solution.

***Example 3 (Mechanical Engineering)***    In dynamic mechanical engineering systems (systems that vary with time) the important variables are position, velocity, acceleration, and force. The equations describing such a system are obtained from free-body diagrams and D'Alembert's principle or Newton's

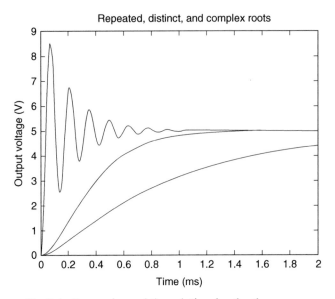

Fig. 1.4. Comparison of the solution for the three cases.

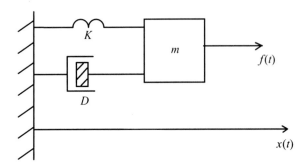

**Fig. 1.5.** Dynamic model of a shock absorber.

laws. An example of a possible automobile shock absorber system is shown in Figure 1.5. The mass of the upper part of the automobile is represented by the block containing $m$, the spring connected between the axle and the body is represented by $K$, and the shock absorber is represented by $D$. The differential equation relating the position of the upper part of the automobile, $x(t)$, to the force applied is

$$\frac{d^2x(t)}{dt^2} + \frac{D}{m}\frac{dx(t)}{dt} + \frac{K}{m} = \frac{1}{m}f(t)$$

When the automobile strikes a bump in the road, it is given an initial displacement and initial velocity upward. This is again a second-order, linear, constant-coefficient ordinary differential equation that we study in Chapter 4. It is also identical in form to that of the electric circuit investigated previously. To provide a smooth ride for passengers, the vehicle position should return rapidly and smoothly to the original position with minimal oscillatory movement about the steady-state position. Hence, the general forms of the solution for the position $x(t)$ are very similar to the electrical circuit problem. The exponential parts of the solution will again have three forms, depending on whether the roots of the characteristic equation

$$p^2 + \frac{D}{m}p + \frac{K}{m} = (p - p_1)(p - p_2) = 0$$

which are

$$p_1, p_2 = -\frac{D}{2m} \pm \frac{1}{2}\sqrt{\left(\frac{D}{m}\right)^2 - 4\frac{K}{m}}$$

are real and distinct, real repeated, or complex:

Case I: roots real and distinct:

$$p_1 \neq p_2 \quad \text{if} \left(\frac{D}{m}\right)^2 > 4\frac{K}{m}$$

Case II: roots real and equal:

$$p_1 = p_2 = p \quad \text{if} \left(\frac{D}{m}\right)^2 = 4\frac{K}{m}$$

Case III: roots complex:

$$p_1, p_2 = \alpha \pm j\beta \quad \text{if} \left(\frac{D}{m}\right)^2 < 4\frac{K}{m}$$

Once again, for the vehicle body to return smoothly and rapidly to its original position, with no oscillation, the desired case is case II for repeated roots, from which we can determine the desired relation between $K$, $D$, and $m$.

**Example 4 (Mechanical Engineering)** Mechanical engineers are frequently concerned with vibrations of beams, rods, and tension cables. Strong oscillations can destroy a system. To describe the general problem, we focus on the vibrating string problem. A string or cable has its two endpoints held fixed, and it is given an initial upward displacement and velocity of displacement as shown in Figure 1.6. The variable $u(x, t)$ gives the vertical position of the string

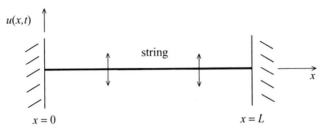

**Fig. 1.6.** The vibrating string problem.

as a function of time $t$ and position along the string $x$. This problem is governed by the "wave equation":

$$\frac{\partial^2 u(x,t)}{\partial x^2} = \frac{1}{v^2}\frac{\partial^2 u(x,t)}{\partial t^2}$$

This is a linear, constant-coefficient, partial differential equation (PDE) whose solution process we study in Chapter 6. Partial derivatives are required here because the displacement of the string $u$ is a function of two independent variables, $t$ (time) and $x$ (position along the string). The general solution to this PDE is of the form

$$u(x,t) = f^+(x - vt) + f^-(x + vt)$$

The functions $f^+(x - vt)$ and $f^-(x + vt)$ are to be determined by the boundary conditions and are said to be "traveling waves" in the same fashion as when two people hold the ends of a jump rope and one gives a rapid snap to that end to the rope. The function $f^+(x - vt)$ is a wave traveling in the $+x$ direction, and the function $f^-(x + vt)$ is a wave traveling in the $-x$ direction. Observe that the solutions can only be a function of $x$, $v$, and $t$ as $x - vt$ and $x + vt$. The combination of these two oppositely traveling waves results in a "standing wave" such as that seen in a vibrating violin string.

***Example 5 (Civil Engineering)***    One of the many design problems that civil engineers encounter is the design of bridge trusses. A simple such structure is shown in Figure 1.7. The stresses in each member shown as vectors $T_1$, $T_2$, and $T_3$ are to be determined. The structure is symmetric about its middle so that the stress vectors in the members of the right half of the structure are not shown but are the same as those in the corresponding members of the left half of the structure. A car is stopped at the middle of the bridge and exerts a force $F$ downward at that point. The structure is assumed to be in equilibrium. Summing the vertical force components and summing the horizontal force components at points $A$, $B$, and $C$ and using trigonometry (Chapter 2) gives four equations:

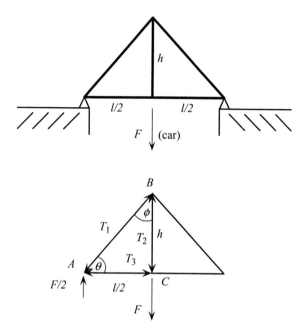

**Fig. 1.7.** Truss design.

$$-T_1 \sin\theta + \frac{F}{2} = 0$$

$$-T_1 \cos\theta - T_3 = 0$$

$$2T_1 \cos\phi + T_2 = 0$$

$$-T_2 - F = 0$$

The dimensions of the truss members, $h$ and $l$, are known, as is the force exerted by the car's weight, $F$, so there are five unknowns: $T_1$, $T_2$, $T_3$, $\theta$, and $\phi$. From trigonometry (Chapter 2) we see that $\theta = \tan^{-1}(h/(l/2))$, $\phi = \tan^{-1}((l/2)/h)$, and $\theta + \phi = 90°$. We develop the following trigonometric identity in Chapter 2:

$$\cos\phi = \cos(90° - \theta)$$
$$= \sin\theta$$

Substituting this to eliminate $\phi$ and eliminating redundant equations gives two equations:

$$2T_1 \sin\theta = F$$

$$T_1 \cos\theta + T_3 = 0$$

Therefore, there are only two unknowns, $T_1$ and $T_3$. Arranging the equations above in matrix form gives

$$\begin{bmatrix} 2\sin\theta & 0 \\ \cos\theta & 1 \end{bmatrix}\begin{bmatrix} T_1 \\ T_3 \end{bmatrix} = \begin{bmatrix} F \\ 0 \end{bmatrix}$$

which are simultaneous, linear, algebraic equations that can be solved by the methods of Chapter 3. These simple equations are rather easy to solve by substitution, giving $T_1 = F/2\sin\theta$ (a tensile stress) and $T_3 = -T_1\cos\theta$ (a compressive stress). But more complicated structures having many beams result in large systems of simultaneous equations that require systematic solution methods which we study in Chapter 3.

***Example 6 (Discrete-Time Systems and the Approximate Solution of Differential Equations)***    The digital technology of today allows powerful computing and data processing. Analog-to-digital converters (A-D converters) allow analog signals to be converted to digital data, which can be processed by computers much faster than by the analog processing used generally a few years ago. Similarly, digital computers can provide rapid but approximate solutions to differential equations (even nonlinear ones, most of which are virtually impossible to solve by hand) by discretizing the variables into discrete increments. For example, consider the first-order ordinary differential equation

$$\frac{dy(t)}{dt} + ay(t) = K$$

where $a$ and $K$ are known. The general solution to this is determined in Chapter 4 subject to the known initial condition $y(0)$ as

$$y(t) = \left[y(0) - \frac{K}{a}\right]e^{-at} + \frac{K}{a}$$

Instead, let us solve this by breaking the $t$ (time) axis into equal increments $\Delta t$ and approximate the derivative as

$$\frac{dy(t)}{dt} \cong \frac{y(t+\Delta t) - y(t)}{\Delta t}$$

Substituting this into the original differential equation gives the "recursion relation"

$$y_{n+1} = (1 - a\Delta t)y_n + \Delta t\, K$$

where we have denoted the value of $y$ at each of the discrete times $t_n = n\Delta t$ as

$$y_n \equiv y(n\Delta t)$$

We can solve these *recursively* as

$$y_1 = (1 - a\,\Delta t)\,y_0 + \Delta t\,K$$

$$y_2 = (1 - a\,\Delta t)\,y_1 + \Delta t\,K$$

$$y_3 = (1 - a\,\Delta t)\,y_2 + \Delta t\,K$$

$$\vdots$$

Solving the first equation [$a$, $y_0 = y(0)$, $\Delta t$, and $K$ are known] gives $y_1$. Substituting this into the second equation, we can solve for $y_2$, and so on. Thus, we can arrive at an approximation in a "marching in time" fashion. But to achieve sufficient accuracy, the time increment $\Delta t$ must be chosen small enough. Hence, we may need very many calculations to reach the solution at a desired time. This recursion relation can be written as a difference equation (Chapter 5) by placing the unknowns on the left and the known quantities on the right:

$$y_{n+1} - (1 - a\,\Delta t)\,y_n = \Delta t\,K$$

In Chapter 5 we determine how to obtain a closed-form equation for the answer *at any time increment*:

$$y_n = \left(y_0 - \frac{K}{a}\right)(1 - a\,\Delta t)^n + \frac{K}{a}$$

*without having to go through the iterative process* (i.e., by going directly to the answer)!

The examples above have shown two important things. First, the equations governing engineering systems fall into a few general categories (simultaneous algebraic equations, ordinary differential equations, difference equations, and partial differential equations). To design engineering systems to accomplish a specific task, it is imperative that you know how to solve these types of equations, and because of the frequency with which you will encounter them, you must be able to recognize them immediately and must have immediate recall of that solution ability. Otherwise, you will have little hope of becoming competent engineers. Second, there are a number of miscellaneous math skills, such as trigonometry and logarithms, that are involved routinely in their solution, and the interpretation of that solution (which is critical to your obtaining "engineering insight" into how engineering systems behave). This book is dedicated to giving you those math skills. If you wish to become a competent engineer, you must learn the essential math skills of this book and have immediate recall of them. If you do, you will proceed successfully toward your goal of becoming a competent engineer. With a mastery of the math skills in this book, anyone can become a competent engineer. Without this mastery, it is doubtful that you will achieve your goal.

When you study this book it is important that you adopt the following attitude about learning the material.

*DO NOT take the attitude that you only need to memorize the equations and their solution. If you do, you will be no better off than when you started since that mode of thinking will not provide you with the long-term ability to use these important math skills and they will be quickly forgotten. Try to understand, in simple terms, what the equations and their solution are "trying to tell you" and why that makes sense. This is how you will develop engineering "insight" and become successful engineers. I will not "prove" the results in the traditional mathematical meaning of a "rigorous" proof. Instead, I will provide you with simple, "commonsense" explanations of the logical meaning of the equations and their solution, as well as simple and "commonsense" methods for obtaining their solution. If you try to think through my commonsense explanations of the solution process, you should be able to extend these results to obtain other useful results on your own. If you adopt this attitude of study, you will achieve the desired long-term retention of these math skills and will be able to use them effectively in your daily study or work.*

I will place a box around certain important results. This is not intended to mean that the result is something you should memorize but is something you should focus on.

# 2 Miscellaneous Math Skills

This chapter covers miscellaneous mathematical topics that are commonly encountered in engineering. Many of these are also common to the math skills of the remaining chapters. The topics are:

- equations of lines, planes, and circles
- areas and volumes of common shapes
- roots of a quadratic equation
- logarithms
- reduction of fractions and lowest common denominators
- long division
- trigonometry
- complex numbers and complex algebra
- common derivatives
- common integrals
- numerical integration

## 2.1 EQUATIONS OF LINES, PLANES, AND CIRCLES

In engineering, functions of one variable, such as $y(x)$ and $f(t)$, are encountered very frequently. Generally, our task is to determine that function as a solution to the governing equations. Once we obtain that solution function, the final task in design is to plot it and thereby begin to understand what it is telling us about the behavior of the system. That is how we obtain the "engineering intuition" to become successful engineers. Generally, that function represents a nonlinear curve. The function $y(x)$ represents a graph in two-dimensional space with the vertical axis as $y$ and the horizontal axis as $x$. Nonlinear curves complicate the interpretation and/or solution of a problem. Hence, it is very common in engineering to make a "piecewise-linear" approximation to the actual curve to facilitate an understanding and/or solution of the problem. A piecewise-linear approximation is a sequence of straight-line segments.

*Essential Math Skills for Engineers*, By Clayton R. Paul
Copyright © 2009 John Wiley & Sons, Inc.

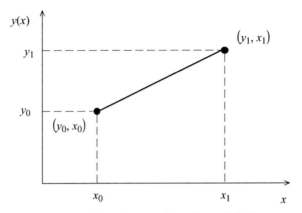

**Fig. 2.1.** Writing the equation of a straight line.

So the first task is to determine the general form of the equation of a straight line (in two-dimensional space) that is shown in Figure 2.1. Note that the dimension axes for a two-dimensional space, $y$ and $x$, are perpendicular to each other. This is said to be a *rectangular coordinate system*. The general form of the equation of this straight line is

$$\boxed{y = mx + b} \qquad \text{straight line} \qquad (2.1)$$

The constants $m$ and $b$ are to be determined for the particular line. The *slope* of the line (which can be positive or negative) is determined from

$$\boxed{m = \frac{y_1 - y_0}{x_1 - x_0}} \qquad (2.2)$$

The constant $b$ is obtained by evaluating (2.1) at either of the two endpoints:

$$\boxed{b = y_1 - mx_1 = y_0 - mx_0} \qquad (2.3)$$

The alternative form of the equation of a straight line is

$$\boxed{y - y_0 = m(x - x_0)} \qquad (2.4a)$$

or

$$\boxed{y - y_1 = m(x - x_1)} \qquad (2.4b)$$

(Any other pairs of coordinates on that line can also be used to determine $m$ and $b$, but the endpoints are usually easier to use.)

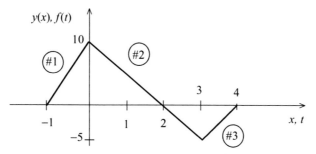

**Fig. 2.2.** A piecewise-linear function.

---

*Example*

Write the equations of the segments of the piecewise-linear function shown in Figure 2.2.

First we write the equation of segment 1. Determine the slope as

$$m = \frac{y_1 - y_0}{x_1 - x_0} = \frac{10 - 0}{0 - (-1)} = 10$$

giving $y = 10x + b$. Then evaluate $b$ as

$$b = y_1 - mx_1 = 10 - 10 \cdot 0 = 10$$

giving the equation of this straight-line segment as $y = 10x + 10$.

---

**A Note About "Sanity Checking" Your Results**

You should "sanity check" all results, meaning that once you have "solved" a problem, you should substitute various values into the supposedly correct equation to see if it indeed is satisfied. I will remind the reader many times throughout the book to "sanity check" a result: Good engineers do this routinely. For example, at the other end of the segment, $x = -1$. Substituting this into the equation gives $y = 0$, which checks.

---

Next we write the equation of segment 2. Determine the slope as

$$m = \frac{y_1 - y_0}{x_1 - x_0} = \frac{(-5) - (10)}{(3) - (0)} = -5$$

giving $y = -5x + b$. Then evaluate $b$ as

$$b = y_1 - mx_1 = (-5) - (-5) \cdot (3) = 10$$

This gives the equation of the segment as $y = -5x + 10$. "Sanity check" this at $x = 0$, which gives $y = 10$, which satisfies the equation for this segment.

Finally, we obtain the equation for segment 3 by first determining the slope as

$$m = \frac{y_1 - y_0}{x_1 - x_0} = \frac{(0) - (-5)}{(4) - (3)} = 5$$

giving $y = 5x + b$. Then determine $b$ as

$$b = y_1 - mx_1 = (0) - (5) \cdot (4) = -20$$

This gives the equation of the segment as $y = 5x - 20$. Again "sanity check" this at $x = 3$, which gives $y = -5$, which satisfies the equation for this segment. So the piecewise-linear representation can be described by

$$y = 10x + 10 \qquad -1 \le x \le 0$$

$$y = -5x + 10 \qquad 0 \le x \le 3$$

$$y = 5x - 20 \qquad 3 \le x \le 4$$

---

Two lines are *parallel* if their slopes are the same (i.e., their $m$'s are the same). Two lines are *perpendicular* (also called orthogonal) if the slope of one is the negative reciprocal of the other. For example, if line 1 is denoted as $y = m_1x + b_1$ and line 2 is denoted as $y = m_2x + b_2$, the two lines are parallel if $m_1 = m_2$ and are perpendicular if $m_1 = -1/m_2$. You should convince yourself of these results by drawing a picture of two lines for these two cases and determining the slopes of each. You can also prove this once we have discussed trigonometry. You should also develop the habit of looking at an equation of a line that is in a more general form and immediately seeing what $m$ and $b$ are for the line. For example, if a line is characterized by $2y + 3x - 5 = 0$, you should immediately recognize that $m = -3/2$ and $b = 5/2$.

The final important result for straight lines is the equation for their length. For straight lines in two-dimensional space, the length is

$$\boxed{\text{length of line} = \sqrt{(x_1 - x_0)^2 - (y_1 - y_0)^2}} \qquad (2.5)$$

The length of segment 2 in Figure 2.2 is

$$\sqrt{(3-0)^2 + (-5-10)^2} = \sqrt{234} = 15.3$$

In a *three-dimensional, rectangular coordinate system*, the axes are denoted as $x$, $y$, and $z$ and are mutually perpendicular. The length of a line whose end coordinates are $(x_0, y_0, z_0)$ and $(x_1, y_1, z_1)$ is

$$\text{length of line} = \sqrt{(x_1 - x_0)^2 + (y_1 - y_0)^2 + (z_1 - z_0)^2} \qquad (2.6)$$

The general equation for a *plane* in three-dimensional space is

$$Ax + By + Cz = D \qquad \text{plane} \qquad (2.7)$$

Observe that this equation is of the same form as the straight line in two-dimensional space, $y = mx + b$, which is a linear equation in the two variables with a constant $b$ added. But note that in a three-dimensional space a linear equation relating the three coordinate axis variables is *not* the equation of a straight line: that relation represents a *plane*. The equation of a plane is also a linear equation in the three variables with a constant $D$ added. In the equation for a line in two-dimensional space, $y = mx + b$, there are two unknowns (the dimension of the space), $m$ and $b$. So it is logical that the equation of a plane should involve only three unknowns (the dimension of the space). If we divide both sides of (2.7) by $D$, we obtain another general equation of a plane:

$$\frac{A}{D}x + \frac{B}{D}y + \frac{C}{D}z = 1 \qquad (2.8)$$

This shows that we only need to determine the three ratios $A/D$, $B/D$, and $C/D$ to determine the equation of a plane. Hence, we need only three sets of points that are in the desired plane, $(x_1, y_1, z_1)$, $(x_2, y_2, z_2)$, and $(x_3, y_3, z_3)$ to evaluate the three unknown ratios, $A/D$, $B/D$, and $C/D$. Evaluating (2.8) at those three points gives

$$x_1(A/D) + y_1(B/D) + z_1(C/D) = 1$$
$$x_2(A/D) + y_2(B/D) + z_2(C/D) = 1$$
$$x_3(A/D) + y_3(B/D) + z_3(C/D) = 1$$

which are three linear, simultaneous equations in the three ratios $A/D$, $B/D$, and $C/D$, whose solution we investigate in Chapter 3. These can also be written in matrix form (Chapter 3) as

$$\begin{bmatrix} x_1 & y_1 & z_1 \\ x_2 & y_2 & z_2 \\ x_3 & y_3 & z_3 \end{bmatrix} \begin{bmatrix} A/D \\ B/D \\ C/D \end{bmatrix} = \begin{bmatrix} 1 \\ 1 \\ 1 \end{bmatrix}$$

---

### Example

Determine the equation of a plane in three-dimensional space containing the three points $(1, 0, 2)$, $(1, 1, 1)$, and $(-1, 1, 0)$. Evaluate the equation of a plane containing these three points to give

$$1(A/D) + 0(B/D) + 2(C/D) = 1$$
$$1(A/D) + 1(B/D) + 1(C/D) = 1$$
$$-1(A/D) + 1(B/D) + 0(C/D) = 1$$

Solving these simultaneous, linear, algebraic equations (by the methods of Chapter 3) gives $A/D = -1/3$, $B/D = 2/3$, and $C/D = 2/3$. So the equation of a plane passing through these three points is

$$-\frac{1}{3}x + \frac{2}{3}y + \frac{2}{3}z = 1$$

or

$$-x + 2y + 2z = 3$$

Hence, we identify $A = -1$, $B = 2$, $C = 2$, and $D = 3$. This equation can also be multiplied by any nonzero number and the equation is not changed. The reader should "sanity check" this result.

---

Now we can investigate determining the equation for a *line* in three-dimensional space. *Two planes in three-dimensional space that are not parallel to each other will intersect in a set of points that determine a line in three-dimensional space.* For example, suppose that two planes are described by $A_1x + B_1y + C_1z = D_1$ and $A_2x + B_2y + C_2z = D_2$. *A line in three-dimensional space is therefore described by the simultaneous equations*

$$\boxed{\begin{aligned} A_1x + B_1y + C_1z &= D_1 \\ A_2x + B_2y + C_2z &= D_2 \end{aligned}} \quad \text{line} \qquad (2.9)$$

Two planes in three-dimensional space described by $A_1x + B_1y + C_1z = D_1$ and $A_2x + B_2y + C_2z = D_2$ are *parallel* if the three coefficients $A$, $B$, and $C$ are proportional by the same constant: $A_1 = kA_2$, $B_1 = kB_2$, and $C_1 = kC_2$, with $k$ nonzero. This is fairly obvious since the coefficients $A$, $B$, and $C$ function like the slope $m$ of a line in two-dimensional space, while the constant $D$ functions like the constant $b$ of a line in two-dimensional space. The constant $D$ basically serves to fix the position of the plane in space. If the planes are parallel, they are the same if $D_1 = kD_2$ but are not the same plane if $D_1 \neq kD_2$. Two planes are *perpendicular* if $A_1A_2 + B_1B_2 + C_1C_2 = 0$. We can demonstrate this result rather easily once we have discussed the topic of trigonometry in Section 2.7. These two criteria for two planes to be parallel or perpendicular can be "sanity checked" by noting that the equation for a plane reduces to that of a line in two-dimensional space if $A = -m$, $B = 1$, $C = 0$, $D = b$ and applying the results for two lines to be parallel or perpendicular. For example, two lines in two-dimensional space described by $y = m_1x + b_1$ and $y = m_2x + b_2$ are parallel if $m_1 = m_2$ and are the same if $b_1 = b_2$. They are perpendicular if $m_1 = -1/m_2$ or, equivalently, $m_1m_2 + 1 = 0$.

The equation of a *circle* in two-dimensional space of radius $r$ centered at $x = x_0$ and $y = y_0$ is

$$(x - x_0)^2 + (y - y_0)^2 = r^2 \qquad \text{circle} \qquad (2.10)$$

Similarly, the equation of a *sphere* in three-dimensional space of radius $r$ and centered at $x = x_0$, $y = y_0$, and $z = z_0$ is

$$(x - x_0)^2 + (y - y_0)^2 + (z - z_0)^2 = r^2 \qquad \text{sphere} \qquad (2.11)$$

## 2.2   AREAS AND VOLUMES OF COMMON SHAPES

The *area* of a *circle* in two-dimensional space of radius $r$ is

$$A = \pi r^2 \qquad \text{area of circle} \qquad (2.12)$$

and the universal constant $\pi$ is

$$\pi = 3.141592653\ldots \qquad (2.13)$$

A *sphere* in three-dimensional space of radius $r$ has a surface *area* of

$$A = 4\pi r^2 \qquad \text{surface area of sphere} \qquad (2.14)$$

and a *volume* of

$$V = \frac{4}{3}\pi r^3 \qquad \text{volume of sphere} \qquad (2.15)$$

A *cylinder* of radius $r$ and length $l$ has a surface *area* of

$$A = 2\pi rl + 2\pi r^2 \qquad \text{surface area of cylinder} \qquad (2.16)$$

and a *volume* of

$$V = \pi r^2 l \qquad \text{volume of cylinder} \qquad (2.17)$$

Both of these results can be obtained easily by visualizing a cylinder as having two circular end caps, each of area $\pi r^2$ and circumference $2\pi r$.

## 2.3   ROOTS OF A QUADRATIC EQUATION

A *quadratic equation* has the form

$$ax^2 + bx + c = 0 \qquad (2.18)$$

The coefficients $a$, $b$, and $c$ are known, and the two values of $x$ that satisfy this equation (the *roots* of the equation), $x_1$ and $x_2$, are to be determined. The roots of this equation are the two values of $x$, $x_1$ and $x_2$, that satisfy it (i.e., which when substituted render the equation zero):

$$\boxed{ax^2 + bx + c = a(x - x_1)(x - x_2) = 0} \qquad (2.19)$$

A general equation for this solution is

$$\boxed{x_1, x_2 = -\frac{b}{2a} \pm \frac{1}{2a}\sqrt{b^2 - 4ac}} \qquad (2.20)$$

It is a common *error* to write the quadratic equation solution as

$$ax^2 + bx + c \neq (x - x_1)(x - x_2)$$

Note that the $a$ is inadvertently missing in this. So *in solving for the roots, it is a good idea to make the leading coefficient of the equation unity* by dividing by $a$ to give $x^2 + b'x + c' = 0$, where $b' = b/a$ and $c' = c/a$, which can be written correctly as

$$x^2 + b'x + c' = (x - x_1)(x - x_2) = 0$$

and whose roots are

$$x_1, x_2 = -\frac{b'}{2} \pm \frac{1}{2}\sqrt{b'^2 - 4c'}$$

The two roots of this equation can be one of three types: (1) real and distinct (not the same) $x_1 = p_1$ and $x_2 = p_2$, where both $p_1$ and $p_2$ are real numbers and $p_1 \neq p_2$, (2) real and repeated (both identical) $x_1 = x_2 = p$ and $p$ is a real number, or (3) complex conjugate, $x_1 = \alpha + j\beta$, $x_2 = \alpha - j\beta$, where $j = \sqrt{-1}$. Complex numbers are discussed in Section 2.8.

There exists an equation for the three roots of a third-order equation such as

$$ax^3 + bx^2 + cx + d = a(x - x_1)(x - x_2)(x - x_3) = 0$$

However, this result is too complicated and used too infrequently to commit to memory or even to look up. A better solution strategy for determining the roots of third- and higher-order equations is to solve them numerically: for example, using your calculator.

## 2.4  LOGARITHMS

Logarithms occur very frequently in engineering problem solutions. The *logarithm of a number to a base b is the power that base must be raised to in order to give the number*:

$$\log_b(x) = n \Rightarrow b^n = x \tag{2.21}$$

There are two common bases. The first base is 10, $b = 10$, and means that

$$\log_{10}(x) = n \Rightarrow 10^n = x \tag{2.22}$$

The second common base is the number *e*:

$$e = 2.71828\ldots \tag{2.23}$$

The logarithm to the base *e* is denoted as

$$\ln(x) = n \Rightarrow e^n = x \tag{2.24}$$

There are three important properties of all logarithms regardless of their base:

$$(a) \; \log(a^m) = m\log(a) \tag{2.25a}$$

$$(b) \; \log(ab) = \log(a) + \log(b) \tag{2.25b}$$

$$(c) \; \log\left(\frac{a}{b}\right) = \log(a) - \log(b) \tag{2.25c}$$

Also observe that if we take the logarithm of the base that is raised to a power, the result is the power:

$$\text{if} \quad 10^x = a \quad \text{then} \quad x\log_{10}(10) = \log_{10}(a) = x$$

and

$$\text{if} \quad e^x = a \quad \text{then} \quad x\ln(e) = \ln(a) = x$$

Finally, it is important to note that the logarithm of the sum of two numbers is *not* the sum of the logarithms of those two numbers:

$$\log(a+b) \neq \log(a) + \log(b)$$

Also,

$$\log_b(1) = 0$$

## 2.5 REDUCTION OF FRACTIONS AND LOWEST COMMON DENOMINATORS

Reduction of several fractions to a lowest common denominator is a skill that seems to have been lost, given the emphasis on the use of calculators in high school. In engineering, it remains a very important required skill.

The sum of two or more fractions should be reduced to a minimal fraction involving the lowest common denominator (LCD). For example:

$$\frac{4}{3} + \frac{1}{4} + \frac{5}{9} = \frac{?}{?}$$

To obtain the LCD, factor each denominator into a product of *prime numbers* (numbers divisible only by 1 and itself: 2, 3, 5, 7, 11, 13, 17, ...) as

$$\frac{4}{3} + \frac{1}{2 \times 2} + \frac{5}{3 \times 3} = \frac{?}{?}$$

The *LCD is the product of the unique factors in the denominators raised to the highest power they appear*, which is $(2 \times 2) \times (3 \times 3) = 36$. Hence, we have

$$\frac{4}{3} + \frac{1}{4} + \frac{5}{9} = \frac{4 \times 12}{36} + \frac{1 \times 9}{36} + \frac{5 \times 4}{36}$$

$$= \frac{77}{36}$$

There are no common factors in this numerator and denominator, so this is its minimal form.

Symbolic fractions are often required to be reduced to the ratio of two polynomials. For example, in solving ordinary differential equations in electric circuit analysis or in the analysis of dynamical systems by the Laplace transform method (Chapter 8), we might obtain

$$F(s) = \frac{N(s)}{D(s)} = \frac{\dfrac{3}{4s^2} - \dfrac{5}{7}}{\dfrac{8}{3s} + \dfrac{6s}{5} - \dfrac{2}{9}}$$

where $s$ is the Laplace transform variable. We must reduce this to the ratio of two polynomials in $s$:

$$F(s) = K \frac{s^m + b_1 s^{m-1} + \cdots + b_m}{s^n + a_1 s^{n-1} + \cdots + a_n}$$

in order to obtain the inverse Laplace transform by expanding $F(s)$ in partial fractions (Chapter 8). To do this we first reduce the numerator, $N(s) = 3/4s^2 - 5/7$, to a LCD. First, ignore the $s$ and determine the LCD of the other denominators, which is $(2 \times 2) \times 7 = 28$. Hence, we write the numerator as

$$N(s) = \frac{3}{4s^2} - \frac{5}{7}$$

$$= \frac{3 \times 7}{28s^2} - \frac{5 \times 4}{28}$$

$$= \frac{21}{28s^2} - \frac{20}{28}$$

Now we take care of the $s^2$ in the denominator by writing

$$N(s) = \frac{21}{28s^2} - \frac{20s^2}{28s^2}$$

$$= \frac{21 - 20s^2}{28s^2}$$

Next we write the denominator in the same fashion. The LCD of the numbers in the denominators of $D(s)$ is $5 \times (3 \times 3) = 45$. So the denominator is written as

$$D(s) = \frac{8}{3s} + \frac{6s}{5} - \frac{2}{9}$$

$$= \frac{8 \times 15}{45s} + \frac{6s \times 9}{45} - \frac{2 \times 5}{45}$$

$$= \frac{120}{45s} + \frac{54s^2}{45s} - \frac{10s}{45s}$$

$$= \frac{120 + 54s^2 - 10s}{45s}$$

Now we write $F(s)$ as

$$F(s) = \frac{\dfrac{21 - 20s^2}{28s^2}}{\dfrac{120 + 54s^2 - 10s}{45s}}$$

$$= \frac{45s}{28s^2} \times \frac{21 - 20s^2}{120 + 54s^2 - 10s}$$

$$= \frac{45s}{28s^2} \times \frac{-20}{54} \times \frac{s^2 - \dfrac{21}{20}}{s^2 - \dfrac{10}{54}s + \dfrac{120}{54}}$$

$$= -\frac{25}{42} \times \frac{s^2 - \dfrac{21}{20}}{s\left(s^2 - \dfrac{10}{54}s + \dfrac{120}{54}\right)}$$

$$= -\frac{25}{42} \times \frac{s^2 - (21/20)}{s^3 - (5/27)s^2 + (20/9)s}$$

giving the final result as the ratio of two polynomials in $s$.

Note that since we have written this as the ratio of two polynomials in $s$, we can factor the numerator and denominator polynomials to write this as

$$F(s) = -\frac{25}{42} \times \frac{(s-z_1)(s-z_2)}{s(s-p_1)(s-p_2)}$$

We can now obtain the roots of the numerator and denominator polynominals using our previous result for the roots of a quadratic equation (which you should check) as $z_1 = j1.025$, $z_2 = -j1.025$, $p_1 = 0.093 + j1.488$, and $p_1 = 0.093 - j1.488$. The denominator also has a root of $s = 0$.

## 2.6  LONG DIVISION

If the ratio of polynomials in $s$ above has a numerator polynomial of order equal to or higher than that of the denominator, we can reduce it using *long division*. For example, consider

$$F(s) = \frac{2s^4 - 3s^2 + s - 1}{3s^2 + 2}$$

To reduce this by long division, we write (all missing powers of $s$ should be included with a zero)

$$
\begin{array}{r}
\frac{2}{3}s^2 - \frac{13}{9} \\
3s^2 + 0s + 2 \overline{) 2s^4 + 0s^3 - 3s^2 + s - 1} \\
2s^4 + 0s^3 + \frac{4}{3}s^2 \\
\hline
\left(-3 - \frac{4}{3}\right)s^2 + s - 1 \\
-\frac{13}{3}s^2 + 0s - \frac{26}{9} \\
\hline
s + \left(-1 + \frac{26}{9}\right) \\
\frac{17}{9}
\end{array}
$$

This gives the expansion as

$$F(s) = \frac{2s^4 - 3s^2 + s - 1}{3s^2 + 2}$$

$$= \frac{2}{3}s^2 - \frac{13}{9} + \frac{s + \frac{17}{9}}{3s^2 + 2}$$

as you can check by multiplying it out.

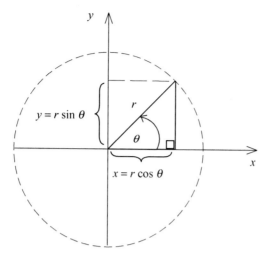

**Fig. 2.3.** The right triangle.

## 2.7  TRIGONOMETRY

### 2.7.1  The Common Trigonometric Functions: Sine, Cosine, and Tangent

The common trigonometric functions, sine (sin), cosine (cos), and tangent (tan), are defined with respect to the right triangle shown in Figure 2.3. The right triangle is inscribed in a circle of radius $r$ that is centered on a rectangular coordinate system consisting of a vertical ($y$) axis and a horizontal ($x$) axis. The angle $\theta$ is measured *counterclockwise* from the horizontal $x$ axis. The included angle between the vertical and horizontal segments in the right triangle is $90°$ and is denoted as a small box. The projection of the radius on the vertical $y$ axis divided by the circle radius $r$ is the sine of the angle $\theta$:

$$\sin \theta = \frac{y}{r}$$

$$(2.26a)$$

The projection of the radius on the horizontal $x$ axis divided by the circle radius $r$ is the cosine of the angle $\theta$:

$$\cos \theta = \frac{x}{r}$$

$$(2.26b)$$

If $90° < \theta < 270°$ ($-270° < \theta < -90°$), the cosine is negative, and if $180° < \theta < 360°$ ($-180° < \theta < 0°$), the sine is negative. The ratio of these projections is the tangent of the angle $\theta$:

$$\tan\theta = \frac{y}{x}$$
$$= \frac{\sin\theta}{\cos\theta}$$

(2.26c)

The hypotenuse of the right triangle (the radius of the circle), $r$, is related to its projections on the $x$ and $y$ axes by the Pythagorean theorem:

$$x^2 + y^2 = r^2$$

(2.27)

Hence, the cosine and sine are related by

$$\cos^2\theta + \sin^2\theta = 1$$

(2.28)

The trigonometric functions can also be represented with an infinite series:

$$\cos\theta = 1 - \frac{\theta^2}{2!} + \frac{\theta^4}{4!} - \frac{\theta^6}{6!} + \cdots$$
$$\approx 1 \quad \text{(small } \theta)$$

(2.29a)

$$\sin\theta = \theta - \frac{\theta^3}{3!} + \frac{\theta^5}{5!} - \frac{\theta^7}{7!} + \cdots$$
$$\approx \theta \quad \text{(small } \theta) \quad (\theta \text{ in radians)}$$

(2.29b)

$$\tan\theta = \frac{\sin\theta}{\cos\theta}$$
$$= \theta + \frac{\theta^3}{3} + \frac{2\theta^5}{15} + \frac{17\theta^7}{315} + \cdots$$
$$\cong \theta \quad \text{(small } \theta)$$

(2.29c)

where the factorial is $n! = 1 \times 2 \times 3 \times \ldots \times n$. Observe the important *small-angle approximations*: $\cos\theta \cong 1$, $\sin\theta \cong \theta$, and $\tan\theta \cong \theta$ for small $\theta$. It is very important to note that the angle $\theta$ *must be in radians* in the three series expansions in (2.29). We are, nevertheless, accustomed to dealing with angles in degrees, particularly on calculators. The conversion between radians and degrees is

$$2\pi\,(\text{radians}) = 360\,(\text{degrees})$$

(2.30)

These infinite series also show some very useful facts about the relationship for negative angles:

$$\cos\theta = \cos(-\theta)$$

(2.31a)

$$\boxed{\sin\theta = -\sin(-\theta)}$$                                   (2.31b)

$$\boxed{\tan\theta = -\tan(-\theta)}$$                                   (2.31c)

You can obtain the angle $\theta$ in terms of the $x$ and $y$ projections with the inverse tangent function on your calculator:

$$\theta = \tan^{-1}\frac{y}{x}$$

But this may give the wrong answer, since calculators put this result into the first or fourth quadrants. To get the correct answer, you must draw a diagram.

---

*Example*

The projections are shown in Figure 2.4. Determine the angle $\theta$.
   Using the inverse tangent function, we obtain, with a calculator,

$$\theta = \tan^{-1}\frac{2}{-3} \neq \underbrace{\tan^{-1}(-0.667)}_{\substack{\text{what you}\\\text{put into}\\\text{your}\\\text{calculator}}} = -33.69°$$

From the diagram, the correct angle (measured counterclockwise from the positive $x$ axis) is

$$\theta = 180° - 33.69° = 146.31°$$

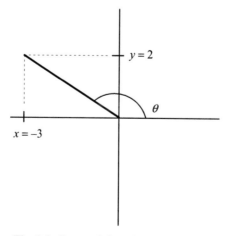

**Fig. 2.4.** Determining the correct angle.

The rectangular-to-polar conversion function for complex numbers on your calculator gives the correct angle.

---

It is very important to be able to visualize plots of these trigonometric functions versus $\theta$. The sine and cosine plots are shown in Figure 2.5. The sine function goes to zero at $\theta = 0°$, $180°$, $360°$ and achieves a maximum of 1 at $\theta = 90°$ and a minimum of $-1$ at $\theta = 270°$. The cosine function goes to zero at $\theta = 90°$, $270°$ and achieves a maximum of 1 at $\theta = 0°$, $360°$ and a minimum of $-1$ at $\theta = 180°$. The tangent plot is shown in Figure 2.6. The tangent function goes to zero at $\theta = 0°$, $180°$, $360°$, and goes to $\pm\infty$ at $\theta = 90°$, $270°$.

There a few other infrequently used types of trigonometric functions: the cosecant: cosec $\theta = 1/\sin\theta$; the secant: sec $\theta = 1/\cos\theta$; and the cotangent: cot $\theta = 1/\tan\theta$.

We can easily write an equation for a sine function in terms of a cosine function having an additional angle, and vice versa, using the ability to trans-

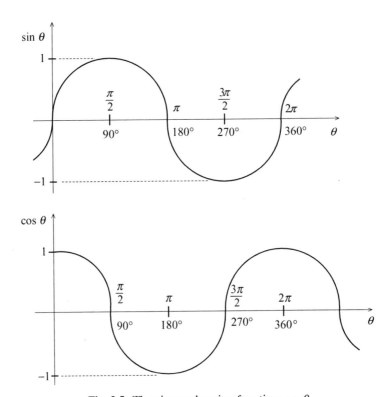

**Fig. 2.5.** The sine and cosine functions vs. $\theta$.

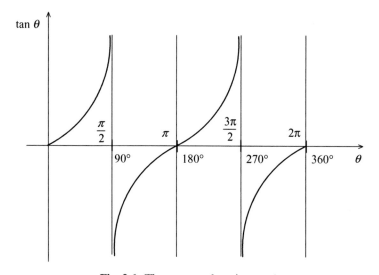

**Fig. 2.6.** The tangent function vs. $\theta$.

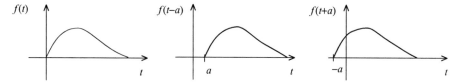

**Fig. 2.7.** Shifting a function.

late or shift a function. This is illustrated in Figure 2.7 for some general function $f(t)$ having an arbitrary shape. We can reason these pictures by choosing a convenient point on the waveform and observing the change in the $t$ value to achieve the shift of that point by a certain amount $a$ left or right. For example, let us choose, arbitrarily, the value of the point where the function is zero: $f(t) = 0$ at $t = 0$. Note that this arbitrarily chosen value of the waveform is located where the argument of the function is zero: $f(0) = 0$. The function $f(t - a)$ is also zero at $f(0)$; in other words, track the movement of the chosen point on the waveform where the argument is zero (i.e., where $t - a = 0$ or at $t = a$). Hence, the waveform $f(t - a)$ is $f(t)$ moved to the *right* by a distance $a$: $f(t - a) = 0$ at $t = a$. Similarly, the function $f(t + a)$ is also zero at $f(0)$; in other words, track the movement of the chosen point on the waveform where the argument is zero (i.e., where $t + a = 0$ or at $t = -a$). Hence, the waveform $f(t + a)$ is $f(t)$ moved to the *left* by a distance $a$: $f(t + a) = 0$ at $t = -a$.

By using this shifting idea for the sine and cosine functions as shown in Figure 2.8, we can obtain the following important relations:

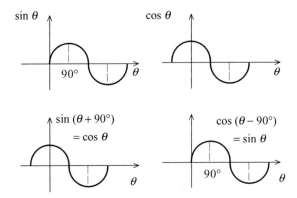

**Fig. 2.8.** Writing the sine in terms of the cosine with an angle, and vice versa.

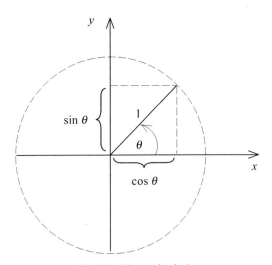

**Fig. 2.9.** The unit circle.

$$\sin(\theta+90°)=\cos\theta \qquad \cos(\theta-90°)=\sin\theta \qquad (2.32a)$$

$$\sin(\theta-90°)=-\cos\theta \qquad \cos(\theta+90°)=-\sin\theta \qquad (2.32b)$$

$$\sin(\theta\pm180°)=-\sin\theta \qquad \cos(\theta\pm180°)=-\cos\theta \qquad (2.32c)$$

which you should verify.

There are important values of $\theta$ for which the reader should easily be able to determine their sine, cosine, and tangent. The first are the cardinal angles of $0°$ ($360°$), $90°$ ($-270°$), $180°$ ($-180°$), and $270°$ ($-90°$). The key to determining these easily is to *draw the unit circle* shown in Figure 2.9, which is Figure

2.3 with a radius of unity, $r = 1$, and observing the projections of this unity radius on the vertical (sin) and horizontal (cos) axes. The values are given in the following table:

| $\theta$ | Sine | Cosine |
|---|---|---|
| 0° (360°) | 0 | 1 |
| 90° (−270°) | 1 | 0 |
| 180° (−180°) | 0 | −1 |
| 270° (−90°) | −1 | 0 |

Results for angles that are multiples of 45° can be found in the same fashion from the unit circle shown in Figure 2.10:

| $\theta$ | Sine | Cosine |
|---|---|---|
| 45° (−315°) | $\dfrac{1}{\sqrt{2}}$ | $\dfrac{1}{\sqrt{2}}$ |
| 135° (−225°) | $\dfrac{1}{\sqrt{2}}$ | $-\dfrac{1}{\sqrt{2}}$ |
| 225° (−135°) | $-\dfrac{1}{\sqrt{2}}$ | $-\dfrac{1}{\sqrt{2}}$ |
| 315° (−45°) | $-\dfrac{1}{\sqrt{2}}$ | $\dfrac{1}{\sqrt{2}}$ |

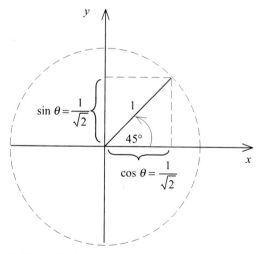

**Fig. 2.10.** The unit circle and angles that are a multiple of 45°.

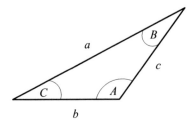

**Fig. 2.11.** A general triangle with sides of length $a$, $b$, and $c$ and angles $A$, $B$, and $C$ opposite the side.

For the sine or cosine of all other angles you should use your calculator, but not for these!

### 2.7.2  Areas of Triangles

Triangles arise quite frequently in engineering, so we need to understand their general properties. Consider the general triangle shown in Figure 2.11. The lengths of the sides are denoted $a$, $b$, and $c$, and the angles opposite that side are denoted $A$, $B$, and $C$.

The first important property of triangles is that the sum of the included angles equals $180°$:

$$\boxed{A + B + C = 180°} \tag{2.33}$$

The *law of sines* relates the ratios of a side and the sine of the angle opposite that side:

$$\boxed{\frac{a}{\sin A} = \frac{b}{\sin B} = \frac{c}{\sin C}} \tag{2.34}$$

The *law of cosines* allows the determination of the length of a side in terms of the lengths of the other sides and their included angle:

$$\boxed{a^2 = b^2 + c^2 - 2bc \cos A} \tag{2.35}$$

"Sanity check" this result by noting that when the triangle is a right triangle with included angle $A = 90°$, $\cos(A = 90°) = 0$ and we obtain the Pythagorean theorem.

The *area* of any triangle, shown in Figure 2.12, whose *height* is $h$ and *base* is $c$ is

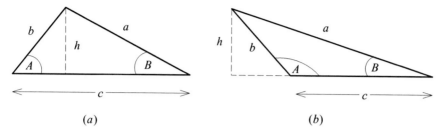

**Fig. 2.12.** The area of a general triangle.

$$area = \frac{1}{2} \times base \times height$$

$$= \frac{1}{2} ch$$

$$= \frac{1}{2} c \underbrace{b \sin A}_{h} = \frac{1}{2} c \underbrace{a \sin B}_{h} = \frac{1}{2} ab \sin C \qquad (2.36)$$

$$= \frac{1}{2} \times (\text{product of two sides}) \times \sin(\text{their included angle})$$

You should "sanity check" the results in (2.36) for both triangles in Figure 2.12 using your knowledge of trigonometry. For example, in the triangle on the left in Figure 2.12(a), the height is obtained from the sides and included angles as $h = b \sin A = a \sin B$. For the triangle on the right in Figure 2.12(b), the height is $h = a \sin B = b \sin(180° - A)$. From (2.32c) $\sin(\theta - 180°) = -\sin \theta$, and (2.31b) shows that $\sin(180° - \theta) = \sin \theta$. Hence, $h = a \sin B = b \sin A$ for that triangle also. For the triangle on the left in Figure 2.12(a) the base is $c = b \cos A + a \cos B$. For the triangle on the right in Figure 2.12(b), the base is $c = a \cos B - b \cos(180° - A)$. From (2.32c), $\cos(\theta - 180°) = -\cos \theta$, and from (2.31a), $\cos(-\theta) = \cos \theta$. Hence, for the triangle on the right in Figure 2.12(b), $c = a \cos B + b \cos A$ also. The results in (2.36) can also be obtained from each other using the law of sines: $b \sin C = c \sin B$, $b \sin A = a \sin B$, and $c \sin A = a \sin C$.

The results in (2.36) include the very common results for the area of a *right triangle* where one of the included angles is 90°, shown in Figure 2.13(a). The triangle on the left in Figure 2.13(a) is a right triangle, which is the triangle on the left in Figure 2.12(a) with an included angle of $B = 90°$. The area of this right triangle is the familiar result $area = (1/2)ac = (1/2)base \times height$ since $\sin B = \sin 90° = 1$. If the triangle on the right in Figure 2.12(b) is a right triangle with an included angle of $A = 90°$, the area of this right triangle is $area = (1/2)bc = (1/2)base \times height$ also.

Similarly, the results in (2.36) include the case of a rectangle with sides $a$ and $c$ and all four angles of 90°, as shown in Figure 2.13(b). The area of the

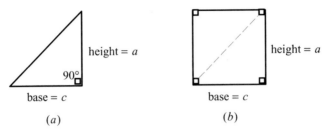

**Fig. 2.13.** Areas of the right triangle and the rectangle.

rectangle can be viewed as being composed of two identical triangles, and hence the area of the rectangle is their sum: *area = ac*.

The general results in (2.36) can also be obtained from the result for a right triangle by viewing the general triangles in Figure 2.12 as being composed of two right triangles and adding the areas of those two triangles. For example, the area of the general triangle in Figure 2.12(a) is *area* = $(1/2)(c-x)h$ + $(1/2)xh$, where the base is the sum of the two bases of the right triangles on the left and right of *h*: $c-x$ and *x*. But this gives the general result *area* = $(1/2)ch$. Similarly, the general triangle in Figure 2.12(b) is the difference of the areas of two triangles, *area* = $(1/2)(c+x)h - (1/2)xh$, where the base of one is $c+x$, and the base of the other is *x*. But this also gives the general result *area* = $(1/2)ch$.

### 2.7.3   The Hyperbolic Trigonometric Functions: Sinh, Cosh, and Tanh

The *hyperbolic trigonometric functions* also arise quite frequently in engineering and are denoted as in the common trigonometric functions earlier but have an *h* appended to those names: the hyperbolic sine (sinh), the hyperbolic cosine (cosh), and the hyperbolic tangent (tanh). The hyperbolic trigonometric functions are defined in terms of the exponential, $e^{ax}$, as

$$\cosh ax = \frac{e^{ax} + e^{-ax}}{2} \tag{2.37a}$$

$$\sinh ax = \frac{e^{ax} - e^{-ax}}{2} \tag{2.37b}$$

and

$$\tanh ax = \frac{\sinh ax}{\cosh ax} = \frac{e^{ax} - e^{-ax}}{e^{ax} + e^{-ax}} \tag{2.37c}$$

The exponential in all these hyperbolic functions has an infinite series expansion of

$$e^{ax} = 1 + \frac{ax}{1!} + \frac{(ax)^2}{2!} + \frac{(ax)^3}{3!} + \cdots \qquad (2.38a)$$

and recall that $e$ is

$$e = 2.71828\ldots \qquad (2.38b)$$

Like the common trigonometric functions, these also have infinite series expansions:

$$\cosh ax = 1 + \frac{(ax)^2}{2!} + \frac{(ax)^4}{4!} + \cdots$$
$$\approx 1 \quad (\text{small } ax) \qquad (2.39a)$$

$$\sinh ax = ax + \frac{(ax)^3}{3!} + \frac{(ax)^5}{5!} + \cdots$$
$$\approx ax \quad (\text{small } ax) \qquad (2.39b)$$

$$\tanh ax = ax - \frac{(ax)^3}{3} + \frac{2(ax)^5}{15} - \frac{17(ax)^7}{315} + \cdots$$
$$\approx ax \quad (\text{small } ax) \qquad (2.39c)$$

You should derive these infinite series by substituting the infinite series expansion of $e^{ax}$ in (2.38a) into their basic definitions given in (2.37a), (2.37b), and (2.37c). There is a simple reason why these series are very similar to those for the common trigonometric functions given in (2.29). We will uncover that reason when we study complex numbers in the next section.

From these infinite series we can obtain some very important results for negative values of the argument:

$$\cosh ax = \cosh(-ax) \qquad (2.40a)$$

$$\sinh ax = -\sinh(-ax) \qquad (2.40b)$$

$$\tanh ax = -\tanh(-ax) \qquad (2.40c)$$

Note that these are identical to the corresponding results for the common trigonometric functions. In addition, you can easily prove the following important identity by substituting (2.37a) and (2.37b):

$$\cosh^2 - \sinh^2 = 1 \qquad (2.41)$$

All of the hyperbolic trigonometric functions involve combinations of the exponential function with positive and negative exponents: $e^{ax}$ and $e^{-ax}$. The exponential $e^{-ax}$ is plotted in Figure 2.14. The exponential with a negative exponent decays to zero for positive $x$ and increases without bound for increasingly negative $x$. Observe that at $x = 0$, $e^{-ax} = 1$, and at $x = 1/a$ it becomes $e^{-1} = 1/e = 0.37$. ... (Recall that $e = 2.71828$. ...) At $x = 5/a$ the exponential has decayed to $e^{-5} = 1/e^5 = 0.0067$. ... At this point the exponential will have decayed sufficiently that it can be considered approximately zero. The positive exponential $e^{ax}$ is this graph but flipped around the vertical axis (i.e., replace $-x$ with $x$). Hence, the positive exponential $e^{ax}$ is 1 at $x = 0$, increases without bound for positive $x$, and decreases to zero for increasingly negative $x$.

The hyperbolic functions sinh $ax$ and cosh $ax$ are plotted in Figure 2.15. You should "sanity check" these by plotting them yourself. *Do not use a calculator.* For example, from (2.37a),

$$\cosh ax = \frac{e^{ax} + e^{-ax}}{2}$$

Therefore, to obtain the graph of cosh $ax$ we simply add the graphs of the two exponentials and divide by 2. At $x = 0$ the sum should be 1, since both exponentials are 1 at $x = 0$. The exponential $e^{-ax}$ increases without bound for negative $x$, as shown in Figure 2.14, thereby dominating the sum, and for positive $x$, $e^{ax}$ also increases without bound, thereby dominating the sum. From (2.37b),

$$\sinh ax = \frac{e^{ax} - e^{-ax}}{2}$$

so that the resulting graph is the difference in the positive and negative exponentials, and hence we obtain the graph shown. At $x = 0$ both exponentials are 1 and the result is zero.

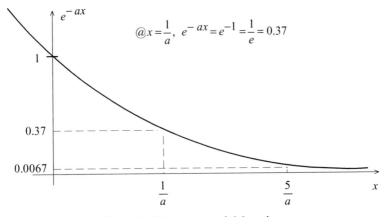

**Fig. 2.14.** The exponential function.

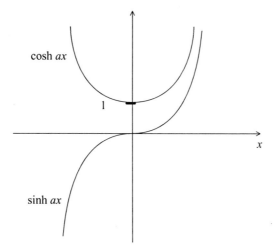

**Fig. 2.15.** Plots of the hyperbolic sine and cosine.

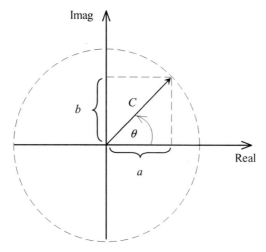

**Fig. 2.16.** Complex numbers.

## 2.8   COMPLEX NUMBERS AND ALGEBRA, AND EULER'S IDENTITY

We will find that the use of complex numbers and complex algebra will markedly simplify the solution of many engineering equations. So we now develop that skill.

A complex number can be thought of as a *vector* in two-dimensional space as illustrated in Figure 2.16. We denote complex numbers in boldface type. A complex number **C** has two parts: a *real part a* and an *imaginary part b*. The real part and the imaginary part are distinguished from one another by the placeholder $j = \sqrt{-1}$:

$$C = a + jb \tag{2.42a}$$

The *real part a* is the projection on the horizontal or real axis, the Real axis, and the *imaginary part b* is the projection on the vertical or imaginary axis, the Imag axis. The representation in (2.42a) is said to be the *rectangular form* of **C**. The complex number can also be written in *polar form* as a magnitude (length of the vector), $C$, and an angle, $\theta$, as

$$C = C \angle \theta \tag{2.42b}$$

The magnitude is denoted using the same symbol as the complex number but without the boldface: $|C| \equiv C$. The magnitude of a complex number is related to the real and imaginary parts again by the Pythagorean theorem:

$$C = \sqrt{a^2 + b^2} \tag{2.43a}$$

and the angle is again measured *counterclockwise* from the positive real axis:

$$\theta = \tan^{-1} \frac{b}{a} \tag{2.43b}$$

The placeholder $j = \sqrt{-1}$ has a number of useful algebraic properties. These are $j^2 = -1$, $j^3 = -\sqrt{-1}$, $j^4 = 1$, and so on. It is important to note that the reciprocal of $j$ is $-j$:

$$\frac{1}{j} = -j$$

The reader can easily prove this by multiplying

$$\frac{1}{j} \times j = 1 = -j \times j$$

There are some special cases that occur for $\theta$ being a multiple of $45°$ that are shown in Figure 2.17. The reader should also show that $a - ja = a\sqrt{2} \angle 315° = a\sqrt{2} \angle -45°$ and $-a - ja = a\sqrt{2} \angle 225° = a\sqrt{2} \angle -135°$. When the angle is some odd multiple of $45°$, the rectangular and polar representations are easy to determine and a calculator should not be used. The real and imaginary parts are equal (although one or both may be negative), and the magnitude is the magnitude of the real or imaginary part multiplied by $\sqrt{2}$.

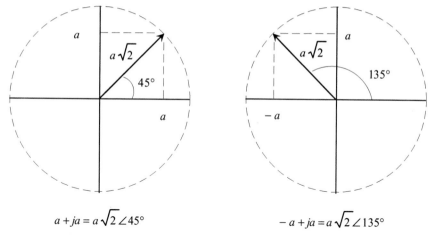

$$a + ja = a\sqrt{2}\angle 45°$$

$$-a + ja = a\sqrt{2}\angle 135°$$

**Fig. 2.17.** Special cases for angles that are a multiple of 45°.

If two complex numbers are denoted as

$$\mathbf{A} = a + jb = A\angle\theta_A \qquad \mathbf{B} = c + jd = B\angle\theta_B \qquad (2.44)$$

*addition, subtraction, multiplication,* and *division* of these two complex numbers are defined as follows. For addition we use the rectangular forms of the numbers, and the result has a real part that is the sum of the real parts and an imaginary part that is the sum of the imaginary parts:

$$\boxed{\mathbf{A} + \mathbf{B} = (a+c) + j(b+d)} \qquad (2.45a)$$

Subtraction is defined similarly:

$$\boxed{\mathbf{A} - \mathbf{B} = (a-c) + j(b-d)} \qquad (2.45b)$$

For multiplication and division we use the polar form:

$$\boxed{\mathbf{AB} = AB\angle(\theta_A + \theta_B)} \qquad (2.45c)$$

and

$$\boxed{\frac{\mathbf{A}}{\mathbf{B}} = \frac{A}{B}\angle(\theta_A - \theta_B)} \qquad (2.45d)$$

Observe that in multiplication, the magnitude of the result is the *product* of the magnitudes, and the angle of the result is the *sum* of the angles. In the case of division, the magnitude of the result is the magnitude of the numerator *divided* by the magnitude of the denominator, and the angle of the result is the angle of the denominator *subtracted* from the angle of the numerator. The rules for addition and subtraction can easily be demonstrated by treating complex numbers like vectors in two-dimensional space. The multiplication and division rules can easily be demonstrated by writing the complex numbers, using Euler's identity (from Section 2.8), as $\mathbf{A} = A\angle\theta_A = Ae^{j\theta_A}$, $\mathbf{B} = B\angle\theta_B = Be^{j\theta_B}$ [e.g., $\mathbf{AB} = Ae^{j\theta_A}Be^{j\theta_B} = ABe^{j(\theta_A+\theta_B)} = AB\angle(\theta_A+\theta_B)$].

---

*Example*

Reduce the following complex function to an equivalent polar form. I should point out that *this is the most frequent and important use of complex algebra in engineering.*

$$F = \frac{(1+j2)+\dfrac{3-j5}{2+j4}}{(1-j3)(4+j1)} = F\angle\theta$$

First, we reduce the numerator:

$$\frac{3-j5}{2+j4} = \frac{5.83\angle-59.04°}{4.47\angle63.43°} = 1.30\angle-122.47° = -0.7 - j1.1$$

Next we reduce the numerator and denominator to polar form, anticipating the division of the numerator and denominator via (2.45d):

$$F = \frac{(1+j2)+(-0.7-j1.1)}{(1-j3)(4+j1)} = \frac{0.3+j0.9}{3.16\angle-71.57°\times4.12\angle14.04°}$$

$$= \frac{0.95\angle71.57°}{13.02\angle-57.53°}$$

$$= 0.07\angle129.10° = -0.04 + j0.05$$

Hence, we identify the result as $\mathbf{F} = 0.07\angle129.10°$.

In all these conversions you should use your calculator but you should "sanity check" all conversions. For example, $3 - j5 = 5.83\angle-59.04°$, which can be "sanity checked" by drawing a simple sketch:

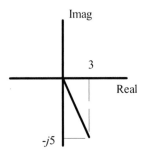

---

There are several other related results. The first is the *conjugate* of a complex number denoted as **C***. It is obtained by replacing $j$ with $-j$ in the rectangular form, or negating the angle in the polar form:

$$\boxed{\mathbf{C^*} = a - jb = C\angle -\theta} \tag{2.46}$$

To obtain the conjugate of a complex equation, simply replace every instance of $j$ with $-j$:

$$\mathbf{F} = \frac{2+j3}{(4+j5)(2e^{j(\pi/3)})} = F\angle\theta \Rightarrow \mathbf{F^*} = \frac{2-j3}{(4-j5)(2e^{-j(\pi/3)})} = F\angle -\theta$$

You should evaluate these expressions and show that $F = 0.282$ and $\theta = -55.03°$. (Euler's identity, discussed next, shows that $2e^{j(\pi/3)} = 2\angle 60°$.) The product of the complex number and its conjugate gives the magnitude squared:

$$\boxed{\begin{aligned} \mathbf{CC^*} &= (a+jb)(a-jb) \\ &= a^2 + b^2 \\ &= Ce^{j\theta}Ce^{-j\theta} \\ &= C\angle\theta C\angle -\theta \\ &= C^2 = |\mathbf{C}|^2 \end{aligned}} \tag{2.47}$$

The square root of a complex number has a magnitude that is the square root of the magnitude and an angle that is one-half the angle:

$$\boxed{\sqrt{\mathbf{C}} = \sqrt{C}\angle\frac{\theta}{2}} \tag{2.48}$$

This can be confirmed by multiplying it times itself giving, by the rules of multiplication in (2.45c),

$$\mathbf{C} = \sqrt{C}\sqrt{C}\angle\left(\frac{\theta}{2}+\frac{\theta}{2}\right) = C\angle\theta$$

Perhaps the most important identity is *Euler's identity* (pronounced "Oiler's" identity). [Use (2.38a) with $ax = j\theta$ and (2.29) to prove this.]

$$\boxed{e^{j\theta} = \cos\theta + j\sin\theta} \tag{2.49}$$

and we can equivalently write the *complex exponential* in polar form as

$$\boxed{e^{j\theta} \equiv 1\angle\theta} \tag{2.50}$$

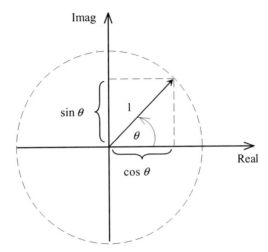

**Fig. 2.18.** The unit circle, demonstrating Euler's identity.

This is evidenced by the unit circle in Figure 2.18. Euler's identity allows us to write the common trigonometric functions of sine and cosine in terms of the complex exponential function as

$$\cos\theta = \frac{e^{j\theta} + e^{-j\theta}}{2} \qquad (2.51a)$$

and

$$\sin\theta = \frac{e^{j\theta} - e^{-j\theta}}{2j} \qquad (2.51b)$$

We can check these results by substituting Euler's identity:

$$\cos\theta = \frac{e^{j\theta} + e^{-j\theta}}{2} = \frac{1}{2}(\cos\theta + j\sin\theta) + \frac{1}{2}(\cos\theta - j\sin\theta)$$

and

$$\sin\theta = \frac{e^{j\theta} - e^{-j\theta}}{2j} = \frac{1}{2j}(\cos\theta + j\sin\theta) - \frac{1}{2j}(\cos\theta - j\sin\theta)$$

which check.

There are numerous important and useful applications of Euler's identity. For example, we can *easily* derive any of the numerous trigonometric identities without having to draw complicated graphs or pictures: Simply substitute Euler's identity and "turn the crank." For example, we can obtain an alternative representation for the product $\sin A \cos B$. To do this we substitute Euler's identity to give

$$\sin A \cos B = \frac{1}{2j}(e^{jA} - e^{-jA})\frac{1}{2}(e^{jB} + e^{-jB})$$

$$= \frac{1}{4j}(e^{j(A+B)} - e^{-j(A+B)} + e^{j(A-B)} - e^{-j(A-B)})$$

$$= \frac{1}{2}\left[\frac{e^{j(A+B)} - e^{-j(A+B)}}{2j}\right] + \frac{1}{2}\left[\frac{e^{j(A-B)} - e^{-j(A-B)}}{2j}\right]$$

$$= \frac{1}{2}\sin(A+B) + \frac{1}{2}\sin(A-B)$$

(Recall that in multiplying two exponential terms, we add their exponents; i.e., $e^a e^b = e^{a+b}$.) By interchanging $A$ and $B$ in this result and noting that $\sin(-\theta) = -\sin\theta$ and $\cos(-\theta) = \cos\theta$, we can also obtain

$$\cos A \sin B = \frac{1}{2}\sin(A+B) - \frac{1}{2}\sin(A-B)$$

Also,

$$\cos A \cos B = \frac{1}{2}(e^{jA} + e^{-jA})\frac{1}{2}(e^{jB} + e^{-jB})$$

$$= \frac{1}{4}(e^{j(A+B)} + e^{-j(A+B)} + e^{j(A-B)} + e^{-j(A-B)})$$

$$= \frac{1}{2}\left[\frac{e^{j(A+B)} + e^{-j(A+B)}}{2}\right] + \frac{1}{2}\left[\frac{e^{j(A-B)} + e^{-j(A-B)}}{2}\right]$$

$$= \frac{1}{2}\cos(A+B) + \frac{1}{2}\cos(A-B)$$

and

$$\sin A \sin B = \frac{1}{2j}(e^{jA} - e^{-jA})\frac{1}{2j}(e^{jB} - e^{-jB})$$

$$= -\frac{1}{4}(e^{j(A+B)} + e^{-j(A+B)} - e^{j(A-B)} - e^{-j(A-B)})$$

$$= -\frac{1}{2}\left[\frac{e^{j(A+B)} + e^{-j(A+B)}}{2}\right] + \frac{1}{2}\left[\frac{e^{j(A-B)} + e^{-j(A-B)}}{2}\right]$$

$$= -\frac{1}{2}\cos(A+B) + \frac{1}{2}\cos(A-B)$$

Observe that in this last identity we have used the property that $j \times j = -1$. Other identities can be obtained in a similar fashion or by adding or subtracting these basic results and recalling that $\sin(-\theta) = -\sin\theta$ and $\cos(-\theta) = \cos\theta$.

$$\sin(A+B) = \sin A \cos B + \cos A \sin B$$

$$\sin(A-B) = \sin A \cos B - \cos A \sin B$$

$$\cos(A+B) = \cos A \cos B - \sin A \sin B$$

$$\cos(A-B) = \cos A \cos B + \sin A \sin B$$

### 2.8.1 Solution of Differential Equations Having Sinusoidal Forcing Functions

We will frequently encounter ordinary differential equations whose right-hand side is a sinusoid:

$$\frac{dV(t)}{dt} + aV(t) = M\cos(\omega t + \theta) \tag{2.52}$$

In this differential equation the terms $a$, $M$, $\omega$, and $\theta$ are known (given), and we want to obtain the equation for $V(t)$ which when substituted into the differential equation satisfies it. A particularly simple way to obtain the solution of this differential equation is to employ Euler's identity. To obtain the solution we replace the right-hand side using Euler's identity with

$$M\cos(\omega t + \theta) \Rightarrow Me^{j(\omega t + \theta)} = M\underbrace{e^{j\theta}}_{1\angle\theta}e^{j\omega t} = (M\angle\theta)e^{j\omega t} \tag{2.53}$$

So we solve instead the differential equation

$$\frac{d\mathbf{V}(t)}{dt} + a\mathbf{V}(t) = M\angle\theta e^{j\omega t} = M\cos(\omega t + \theta) + jM\sin(\omega t + \theta) \tag{2.54a}$$

Since the right-hand side of this equation is now a complex number, the solution to it will also be a complex number having a real and an imaginary component, which we write in rectangular form as

$$\mathbf{V} = V_R + jV_I \tag{2.54b}$$

Substituting this into (2.54a) and collecting terms gives

$$\left[\frac{dV_R(t)}{dt} + aV_R(t)\right] + j\left[\frac{dV_I(t)}{dt} + aV_I(t)\right] = M\cos(\omega t + \theta) + jM\sin(\omega t + \theta)$$

$$\tag{2.54a}$$

Comparing real and imaginary parts on the left and right sides gives two differential equations to be solved:

$$\frac{dV_R(t)}{dt} + aV_R(t) = M\cos(\omega t + \theta) \tag{2.55a}$$

and

$$\frac{dV_I(t)}{dt} + aV_I(t) = M\sin(\omega t + \theta) \qquad (2.55b)$$

The real part of this in (2.55a) is the original differential equation we wanted to solve. So the real part of the solution in (2.54b), $V_R$, is the desired solution.

Is it easier to solve (2.54a) than the original equation in (2.52)? The answer to this is, as we will see, a resounding YES! We assume a *form* of the solution that is of the same *form* as the right-hand side:

$$\mathbf{V}(t) = V\angle\phi e^{j\omega t} \qquad (2.56)$$

and substitute. The derivative of this form with respect to $t$ is very easy to obtain since

$$\frac{d}{dt}e^{j\omega t} = (j\omega)e^{j\omega t} \qquad (2.57)$$

This converts the differential equation in (2.54a) to an algebraic equation:

$$(j\omega + a)V\angle\phi e^{j\omega t} = M\angle\theta e^{j\omega t} \qquad (2.58)$$

Canceling the $e^{j\omega t}$ that is common to both sides and rearranging gives

$$V\angle\phi = \frac{M\angle\theta}{j\omega + a} \qquad (2.59)$$

But this is very easy to reduce and determine $V$ and $\phi$ using complex arithmetic. Hence, the solution to the differential equation in (2.52) is

$$V(t) = V\cos(\omega t + \phi) \qquad (2.60)$$

If the right-hand side of the differential equation in (2.52) had been $M\sin(\omega t + \theta)$, the solution for $V$ and $\phi$ would not change but the solution to the differential equation would simply be

$$V(t) = V\sin(\omega t + \phi) \qquad (2.61)$$

In other words, if the right-hand side of the differential equation is a cosine, the solution is a cosine. If the right-hand side of the differential equation is a sine, the solution is a sine. But both have the same magnitude $V$ and phase angle $\phi$.

## 2.9   COMMON DERIVATIVES AND THEIR INTERPRETATION

There are an unlimited number of functions and their derivatives. Fortunately, only the following ones are used frequently enough to merit committing them to memory. For derivatives of all other functions either derive them or look them up in tables of derivatives.

(1) $\boxed{\dfrac{d}{dx}(u(x)+v(x)) = \dfrac{du}{dx}+\dfrac{dv}{dx}}$

(2) $\boxed{\dfrac{du(x)v(x)}{dx} = u\dfrac{dv}{dx}+v\dfrac{du}{dx}}$

(3) $\boxed{\dfrac{d}{dx}\dfrac{u(x)}{v(x)} = \dfrac{v\dfrac{du}{dx}-u\dfrac{dv}{dx}}{v^2}}$

(4) $\boxed{\dfrac{d}{dx}f(u(x)) = \dfrac{df(u)}{du}\cdot\dfrac{du}{dx}}$

(5) $\boxed{\dfrac{d}{dx}x^n = nx^{n-1}}$

(6) $\boxed{\dfrac{d}{dx}e^{ax} = ae^{ax}}$

(7) $\boxed{\dfrac{d}{dx}\sin ax = a\cos ax}$

(8) $\boxed{\dfrac{d}{dx}\cos ax = -a\sin ax}$

(9) $\boxed{\dfrac{d}{dx}\sinh ax = a\cosh ax}$

(10) $\boxed{\dfrac{d}{dx}\cosh ax = a\sinh ax}$

(11) $\boxed{\dfrac{d}{dx}\ln x = \dfrac{1}{x}}$

The first four are general properties and (2) is called the *chain rule*. The remaining are the only ones that are very commonly encountered in engineering. We can easily determine these from their basic definitions learned previously. The derivative of the exponential in (6) can easily be obtained from the infinite series expansion of the exponential $e^{ax}$ that is given in (2.38a) and using the derivative of powers of $x$ given in (5):

$$\frac{d}{dx}e^{ax} = \frac{d}{dx}\left(1+\frac{ax}{1!}+\frac{(ax)^2}{2!}+\frac{(ax)^3}{3!}+\cdots\right)$$

$$= \frac{a}{1!}+\frac{2a^2x}{1\times2}+\frac{3a(ax)^2}{1\times2\times3}+\cdots$$

$$= a\left(1+\frac{ax}{1!}+\frac{(ax)^2}{2!}+\frac{(ax)^3}{3!}+\cdots\right)$$

$$= ae^{ax}$$

and we recall that the factorial is defined as $n! = 1\times2\times3\times\ldots\times n$.

The derivatives of the common trigonometric functions in (7) and (8) are easily determined from their expansions using Euler's identity in (2.51) and using property (1) of the derivative of a sum and property (6) above for the derivative of an exponential:

$$\frac{d}{dx}\cos ax = \frac{d}{dx}\left(\frac{e^{jax}+e^{-jax}}{2}\right)$$

$$= ja\left(\frac{e^{jax}-e^{-jax}}{2}\right)$$

$$= -a\left(\frac{e^{jax}-e^{-jax}}{2j}\right)$$

$$= -a\sin ax$$

and

$$\frac{d}{dx}\sin ax = \frac{d}{dx}\left(\frac{e^{jax}-e^{-jax}}{2j}\right)$$

$$= ja\left(\frac{e^{jax}+e^{-jax}}{2j}\right)$$

$$= a\left(\frac{e^{jax}+e^{-jax}}{2}\right)$$

$$= a\cos ax$$

We have used the property of complex numbers that $1/j = -j$ in the first result.

The derivatives of the hyperbolic sine and cosine given in (9) and (10) are easily derived in a similar fashion using the derivative of the exponential in (6) and the definitions of these functions given in (2.37):

$$\frac{d}{dx}\cosh ax = \frac{d}{dx}\left(\frac{e^{ax}+e^{-ax}}{2}\right)$$

$$= a\left(\frac{e^{ax}-e^{-ax}}{2}\right)$$

$$= a\sinh ax$$

and

$$\frac{d}{dx}\sinh ax = \frac{d}{dx}\left(\frac{e^{ax}-e^{-ax}}{2}\right)$$

$$= a\left(\frac{e^{ax}+e^{-ax}}{2}\right)$$

$$= a\cosh ax$$

Perhaps the most important property of the derivative is that

---

*the derivative represents the instantaneous slope of the function.*

---

### *Example*

We see from the discussion above that can plot the derivative of a "piecewise-linear" function such as that shown in Figure 2.19 by plotting the instantaneous slopes of the segments. The slopes are −5 for −1 ≤ $x$ ≤ 0, 15/2 = 7.5 for 0 ≤ $x$ ≤ 2, 0 for 2 ≤ $x$ ≤ 3, and −10 for 3 ≤ $x$ ≤ 4.

---

The derivative has a number of other important uses. We can determine the value of $x$ where the function $f(x)$ achieves a maximum or a minimum as shown in Figure 2.20. Since the points where the function achieves a maximum or a minimum are where the *slope* of the function is zero, we can obtain those points by determining the derivative of the function and solving for the values of $x$ that make this derivative zero:

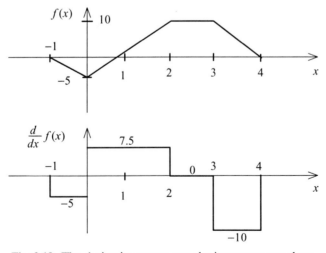

**Fig. 2.19.** The derivative represents the instantaneous slope.

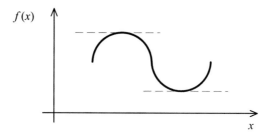

**Fig. 2.20.** Determining points where the function is a maximum or a minimum.

$$\left. \frac{d}{dx}f(x) \right|_{x=a} = 0 \quad \text{a maximum or minimum may occur at } a \quad (2.62)$$

A function $f(x)$ may (and usually will) have several points where the slope is zero. But only one of these will be the maximum or minimum value of the entire function. To determine the actual point where the function is a maximum or a minimum, we solve for all the values of $x$ where the derivative is zero and evaluate the function at each of these to determine which one is the location of the maximum and which one is the location of the minimum for the entire function.

It occurs quite frequently that when we evaluate the limit of the ratio of two functions, this limit is indeterminant. The true limit (if one actually exits) can be determined from l'Hôpital's rule (pronounced as "lowpitaal"), where we differentiate the numerator and differentiate the denominator and reevaluate the limit of this ratio:

$$\text{If } \lim_{x \to a} \frac{f(x)}{g(x)} = \frac{0}{0} \text{ or } \frac{\infty}{\infty} \quad \text{then } \lim_{x \to a} \frac{f(x)}{g(x)} = \lim_{x \to a} \frac{f'(x)}{g'(x)} \quad (2.63)$$

where $f'(x) = \dfrac{d}{dx}f(x)$. If this limit is indeterminate, we can continue to differentiate the numerator and denominator and evaluate the limit until a result is obtained (if it exists for this ratio of functions).

## 2.10   COMMON INTEGRALS AND THEIR INTERPRETATION

As was the case for derivatives, there are an unlimited number of integrals of functions, but only a very few (noted below) are encountered frequently enough in engineering to merit their memorization.

(1) $\boxed{\int (u+v)\,dx = \int u\,dx + \int v\,dx}$ 　　(2) $\boxed{\int u\,dv = uv - \int v\,du}$

(3) $\boxed{\int x^n\,dx = \dfrac{x^{n+1}}{n+1}}$ 　　(4) $\boxed{\int e^{ax}\,dx = \dfrac{1}{a}e^{ax}}$

(5) $\boxed{\int \sin ax\,dx = -\dfrac{1}{a}\cos ax}$ 　　(6) $\boxed{\int \cos ax\,dx = \dfrac{1}{a}\sin ax}$

(7) $\boxed{\int \sinh ax\,dx = \dfrac{1}{a}\cosh ax}$ 　　(8) $\boxed{\int \cosh ax\,dx = \dfrac{1}{a}\sinh ax}$

(9) $\boxed{\int \dfrac{1}{x}\,dx = \ln|x|}$

All other integrals are too infrequently encountered in engineering and are too difficult to remember. For all other integrals, get a good table of integrals, such as

H. B. Dwight, *Tables of Integrals and Other Mathematical Data*, fourth edition, Macmillan, New York, 1961.

You should "sanity check" the results in (3)–(9) by differentiating each result and showing that it gives the integrand. For example, doing this for (3) gives

$$\frac{d}{dx}\left(\frac{x^{n+1}}{n+1}\right) = (n+1)\frac{x^n}{n+1} = x^n$$

The integral of the exponential in (4) can be derived from the infinite series expansion of the exponential $e^{ax}$ that is given in (2.38a) and using the integral of powers of $x$ given in (3):

$$\int e^{ax}\,dx = \int\left(e^{ax} = 1 + \frac{ax}{1!} + \frac{(ax)^2}{2!} + \frac{(ax)^3}{3!} + \cdots\right)dx$$

$$= x + \frac{ax^2}{2\times1!} + \frac{a^2 x^3}{3\times2!} + \frac{a^3 x^4}{4\times3!} + \cdots$$

$$= \frac{1}{a}\left(ax + \frac{a^2 x^2}{2!} + \frac{a^3 x^3}{3!} + \frac{a^4 x^4}{4!} + \cdots\right)$$

$$= \frac{1}{a}e^{ax}$$

The integrals of the common trigonometric functions in (5) and (6) are easily determined from their expansions using Euler's identity in (2.51), using property (1) of the integral of a sum and property (4) above for the integral of an exponential:

$$\int \cos ax \, dx = \int \left( \frac{e^{jax} + e^{-jax}}{2} \right) dx$$

$$= \frac{1}{ja} \left( \frac{e^{jax} - e^{-jax}}{2} \right)$$

$$= \frac{1}{a} \sin ax$$

and

$$\int \sin ax \, dx = \int \left( \frac{e^{jax} - e^{-jax}}{2j} \right) dx$$

$$= \frac{1}{ja} \left( \frac{e^{jax} + e^{-jax}}{2j} \right)$$

$$= -\frac{1}{a} \cos ax$$

and we have used $j \times j = -1$ in the last result.

The integrals of the hyperbolic sine and cosine given in (7) and (8) are easily derived in a similar fashion using the integral of the exponential in (4) and the definitions of these functions given in (2.37):

$$\int \cosh ax \, dx = \int \left( \frac{e^{ax} + e^{-ax}}{2} \right) dx$$

$$= \frac{1}{a} \left( \frac{e^{ax} - e^{-ax}}{2} \right)$$

$$= \frac{1}{a} \sinh ax$$

and

$$\int \sinh ax \, dx = \int \left( \frac{e^{ax} - e^{-ax}}{2} \right) dx$$

$$= \frac{1}{a} \left( \frac{e^{ax} + e^{-ax}}{2} \right)$$

$$= \frac{1}{a} \cosh ax$$

As a check it is important to note that *if we differentiate the result of an integration, it should give back the integrand.* For example, to check the integral of the sine in (5), $\int \sin ax \, dx = -(1/a)\cos ax$, differentiate the result as

$$\frac{d}{dx}\left(-\frac{1}{a}\cos ax\right) = \sin ax$$

using the property of the derivative of a cosine:

$$\frac{d}{dx}\cos ax = -a\sin ax$$

The table of integrals above has no limits of integration on the integrals. To be useful we must integrate between two extremes of $x$: $\int_a^b f(x)dx$. It is very important to observe that

---

$\int_a^a f(x)dx$ represents the area underneath the function between $x = a$ and $x = b$.

---

The area above the $x$ axis represents a positive area, and area below the $x$ axis represents a negative area.

---

### Example

For example, the "piecewise-linear" function $f(t)$ shown in Figure 2.21 is to be integrated from $t = -\infty$ to $t$: $\int_{-\infty}^t f(\tau)d\tau$. This simply says that we want to obtain

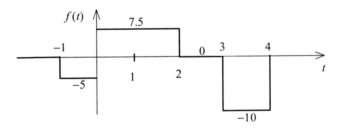

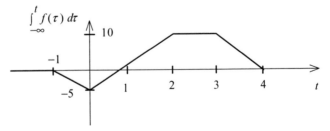

**Fig. 2.21.** The integral as the accumulated area under a curve.

the *accumulated area* contained under the curve from $t = -\infty$ to some present time $t$. Observe that we have used the dummy integral $\tau$ in the integrand so as not to confuse it with the upper limit $t$ for this particular integral. When we write the integral as $\int_{-\infty}^{t} f(\tau)d\tau$, this means that we start accumulating the area at $t = -1$ [the curve for $f(t)$ in Figure 2.21 is zero for $t < -1$], and for each time $t$, we stop, accumulate the area, and write it on the graph for the result at that $t$. For piecewise-linear graphs this is easy to do. Since the segments are straight lines (in this case, they are horizontal lines with zero slope) we only need to determine the accumulated areas at the ends of the line segments, $t = 0, t = 2$, $t = 3, t = 4$, which are $-5, -5 + 15 = 10, 10 + 0 = 10$, and $10 - 10 = 0$, respectively, and we have used the basic result that the area of a rectangle with sides $a$ and $b$ is *area* = $ab$. Then draw straight lines between the accumulated values. We can draw straight lines between these accumulated values because the function $f(t)$ consists of horizontal segments with zero slope and hence the area is accumulated linearly for this function shape.

---

Some integrals do not fit into the simple forms above and are therefore difficult to integrate. A very useful technique for these integrals is to reduce them to a known form using a *change of variables*.

---

### Example

Suppose that we wish to evaluate the integral

$$\int_{2}^{3} \frac{1}{10-x} dx = ?$$

To put this into a form whose integral we know, we make a *change of variables*:

$$\lambda = 10 - x$$

so that

$$d\lambda = -dx$$

The limits of the integral become

$$x = 2 \Rightarrow \lambda = 8$$
$$x = 3 \Rightarrow \lambda = 7$$

Substituting (1) the change of variables in the integrand and (2) the change of limits gives

$$\int_2^3 \frac{1}{10-x}\,dx = -\int_8^7 \frac{1}{\lambda}\,d\lambda$$
$$= \int_7^8 \frac{1}{\lambda}\,d\lambda$$
$$= \ln(8) - \ln(7)$$

---

## 2.11   NUMERICAL INTEGRATION

Despite the simplifications above, there are integrals encountered in engineering that cannot be evaluated in "closed form" (i.e., resulting in an equation for their value). A very useful technique for evaluating these integrals approximately is to segment the integrand function, approximate the areas of the function over each of these segments, and add the results. For example, suppose that we wish to evaluate the integral $\int_a^b f(t)\,dt$. We divide the horizontal axis into equal-length segments $\Delta t$, approximate the curve over these segments in a fashion that permits an immediate determination of the area over this segment, and add the resulting areas. For example, Figure 2.22 shows approximating a function with rectangles. The approximate area under the curve is

$$\int_{t_0}^{t_n} y(t)\,dt \cong y(t_0)\Delta t + y(t_1)\Delta t + y(t_3)\Delta t + \cdots$$
$$= [y(t_0) + y(t_1) + y(t_3)\Delta t + \cdots]\Delta t$$

(2.64)

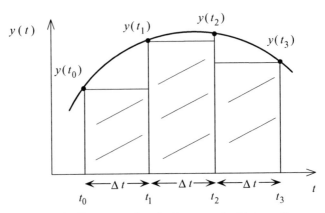

**Fig. 2.22.** The rectangle rule for numerical integration.

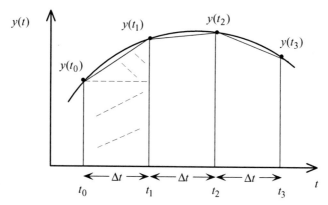

**Fig. 2.23.** The trapezoidal rule for numerical integration.

This is called the *rectangle rule* for numerical integration.

A better approximation is the *trapezoidal rule*, where the horizontal axis is again broken into equal-length segments $\Delta t$, and the area under the curve over these segments is approximated by *trapezoids* as shown in Figure 2.23. The area under the first segment is the sum of the area of a rectangle and the area of a triangle:

$$y(t_0)\Delta t + \frac{1}{2}[y(t_1) - y(t_0)]\Delta t = \frac{1}{2}[y(t_0) + y(t_1)]\Delta t$$

This result can be interpreted either as (1) the area of a rectangle whose height is the *average* of the vertical coordinates or (2) the area of a triangle whose height is the *sum* of the vertical coordinates. The total area from $t_0$ to $t_n$ for $n$ segments is approximately

$$\int_{t_0}^{t_n} y(t)\,dt \cong \frac{1}{2}[y(t_0) + y(t_1)]\Delta t + \frac{1}{2}[y(t_1) + y(t_2)]\Delta t + \cdots + \frac{1}{2}[y(t_{n-1}) + y(t_n)]\Delta t$$

$$= \left[\frac{1}{2}y(t_0) + y(t_1) + y(t_2) + \cdots + y(t_{n-1}) + \frac{1}{2}y(t_n)\right]\Delta t$$

(2.65)

This called the *trapezoidal rule* for numerical integration. There are other more accurate (but more complicated) rules for numerical integration, but they also rely on dividing the horizontal axis into discrete segments.

# 3 Solution of Simultaneous, Linear, Algebraic Equations

In the examples in Chapter 1 we saw that perhaps the most frequently encountered and important math skill for engineers is the solution of simultaneous, linear, algebraic equations. This chapter is devoted to studying methods for their solution.

## 3.1 HOW TO IDENTIFY SIMULTANEOUS, LINEAR, ALGEBRAIC EQUATIONS

What do they look like? Two such simultaneous equations are in the form

$$\left.\begin{array}{l} a_{11}x_1 + a_{12}x_2 = b_1 \\ a_{21}x_1 + a_{22}x_2 = b_2 \end{array}\right| \tag{3.1a}$$

The *two* unknowns to be solved for are denoted as $x_1$ and $x_2$. The coefficients $a_{ij}$ are *known* (given), and the first subscript $i$ denotes the equation, while the second subscript $j$ denotes the unknown it multiplies. The two terms on the right-hand side, $b_1$ and $b_2$, are *known* (given). Three such simultaneous equations are of the form

$$\left.\begin{array}{l} a_{11}x_1 + a_{12}x_2 + a_{13}x_3 = b_1 \\ a_{21}x_1 + a_{22}x_2 + a_{23}x_3 = b_2 \\ a_{31}x_1 + a_{32}x_2 + a_{33}x_3 = b_3 \end{array}\right| \tag{3.1b}$$

We will not be concerned with more that three equations since those are too difficult to be solved by hand: They are best solved with a computer or your calculator. But we will also be concerned with two and three equations where the coefficients, $a_{ij}$, and the right-hand-side items, $b_i$, are in terms of symbols (i.e., are *literal* equations). Solution of these equations is considerably more difficult when the coefficients are given in terms of symbols.

---

*Essential Math Skills for Engineers*, By Clayton R. Paul
Copyright © 2009 John Wiley & Sons, Inc.

How do we identify these? There are three terms in their classification: *simultaneous*, *linear*, and *algebraic*. The term *simultaneous* clearly means that there is more than one equation, and for example, for the case of two equations in (3.1a), $x_1$ and $x_2$ must satisfy both equations *simultaneously*. The term *algebraic* means that there are no derivatives of the $x$'s. There are two general types of equations: algebraic and differential. A differential equation would contain at least one derivative, such as

$$\frac{dx(t)}{dt} + ax(t) = b(t)$$

We study the solution to differential equations in Chapter 4. Equations are also either linear or nonlinear. A *linear* algebraic equation would have the unknowns, for example $x_1$ and $x_2$, *not* raised to a power greater than 1 and/or *not* appear as products of each other, such as $x_1x_2$. An example of a nonlinear equation is

$$a_{11}x_1x_2 + a_{12}x_2^3 = b_1$$

The equation of a straight line in (2.1) is a linear equation. Nonlinear equations, even algebraic ones, are very difficult to solve, and we do not investigate their solution in this book. Although nonlinear algebraic and differential equations appear in engineering, they are best solved either (1) with a computer using approximate numerical techniques, or (2) by making linear approximations to them so that they can be solved approximately using the methods of this chapter.

There are three *systematic* ways of solving simultaneous, linear, algebraic equations: (1) Cramer's rule, (2) Gauss elimination, and (3) matrix inverse. Cramer's rule is a very simple, straightforward, and systematic method of solving two and three simultaneous equations even if the knowns in these equations, the $a$'s and the $b$'s, are in terms of symbols instead of numbers. For numbers of equations greater that three, the computational burden in using Cramer's rule is prohibitive and the other two methods should be used.

## 3.2    THE MEANING OF A SOLUTION

Two equations in two unknowns in (3.1a) represent two straight lines in two-dimensional space, as illustrated in Figure 3.1. Observe that these two straight lines intersect at a point that represents their simultaneous solution: $X_1$ and $X_2$. Observe also that the intersection of these lines with the horizontal or $x_1$ axis is obtained by setting $x_2 = 0$ (which the horizontal $x_1$ axis represents) and solving each equation for the resulting value of $x_1$: $x_1 = b_1/a_{11}|_{x_2=0}$ and $x_1 = b_2/a_{21}|_{x_2=0}$. Their intersections with the $x_2$ axis is for $x_1 = 0$ (which the vertical $x_2$ axis represents) and solving each equation for the resulting value of $x_2$: $x_2 = b_1/a_{12}|_{x_1=0}$ and $x_2 = b_2/a_{22}|_{x_1=0}$.

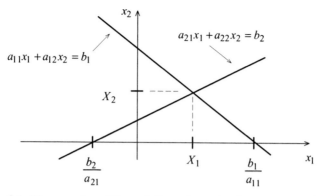

**Fig. 3.1.** The meaning of the solution of two simultaneous equations.

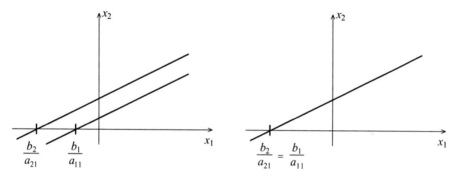

**Fig. 3.2.** No solutions or an infinite number of solutions.

The intersection of the two lines in Figure 3.1 gives the *unique* solutions for the two simultaneous equations. There are two other possible solutions shown in Figure 3.2. If the two lines are parallel but are not the same, there are no intersections and no solutions. If the two lines are coincident, there are an infinite number of solutions.

The solution of three equations in three unknowns represents the intersection of three *planes* in three-dimensional space. The same results for an infinite number of solutions or no solutions for two equations apply to three equations: Two or all of the three planes may be parallel but not coincident, resulting in no solutions, or two or all of the three planes may be coincident, resulting in an infinite number of solutions.

## 3.3   CRAMER'S RULE AND SYMBOLIC EQUATIONS

Cramer's rule is a very systematic method of solving sets of two and three simultaneous algebraic equations even if the coefficients in those equations, the $a$'s, and the right-hand sides, the $b$'s, are symbols rather than numbers. The rule was named after the Swiss mathematician who discovered it.

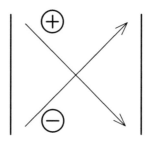

**Fig. 3.3.** Evaluation of a $2 \times 2$ determinant.

First we need to discuss the evaluation of a *determinant*. A determinant is indicated with two vertical bars enclosing the coefficients of the equations. A $2 \times 2$ determinant is evaluated as

$$\begin{vmatrix} a_{11} & a_{12} \\ a_{21} & a_{22} \end{vmatrix} = a_{11}a_{22} - a_{12}a_{21} \tag{3.2}$$

Note that the value of a $2 \times 2$ determinant is the product of the *main diagonal terms*, $a_{11}$ and $a_{22}$, minus the product of the *off-diagonal terms*, $a_{12}$ and $a_{21}$. This is symbolized as shown in Figure 3.3.

A $3 \times 3$ determinant is denoted similarly:

$$\begin{vmatrix} a_{11} & a_{12} & a_{13} \\ a_{21} & a_{22} & a_{23} \\ a_{31} & a_{32} & a_{33} \end{vmatrix} \tag{3.3}$$

The evaluation of a $3 \times 3$ determinant, however, is more complicated than for a $2 \times 2$ determinant. The evaluation of a $3 \times 3$ determinant is accomplished by the following three steps:

1. Choose a row *or* column to *expand* along.
2. Multiply each element in that row or column by $(-1)^{i+j}$, where that element is in the $i$th row and the $j$th column according to the "checkerboard sign pattern":

$$\begin{vmatrix} + & - & + & \cdots \\ - & + & - & \cdots \\ + & - & + & \cdots \\ \vdots & \vdots & \vdots & \ddots \end{vmatrix}$$

and by the determinant of the elements remaining after that element's row and column is struck out.

3. Sum the $n$ quantities obtained in step 2.

---

***Example***

Expand along the first row:

$$\begin{vmatrix} a_{11} & a_{12} & a_{13} \\ a_{21} & a_{22} & a_{23} \\ a_{31} & a_{32} & a_{33} \end{vmatrix} = a_{11}\begin{vmatrix} a_{22} & a_{23} \\ a_{32} & a_{33} \end{vmatrix} - a_{12}\begin{vmatrix} a_{21} & a_{23} \\ a_{31} & a_{33} \end{vmatrix}$$

$$+ a_{13}\begin{vmatrix} a_{21} & a_{22} \\ a_{31} & a_{32} \end{vmatrix}$$

or, expand along the second column:

$$\begin{vmatrix} a_{11} & a_{12} & a_{13} \\ a_{21} & a_{22} & a_{23} \\ a_{31} & a_{32} & a_{33} \end{vmatrix} = -a_{12}\begin{vmatrix} a_{21} & a_{23} \\ a_{31} & a_{33} \end{vmatrix} + a_{22}\begin{vmatrix} a_{11} & a_{13} \\ a_{31} & a_{33} \end{vmatrix}$$

$$- a_{32}\begin{vmatrix} a_{11} & a_{13} \\ a_{21} & a_{23} \end{vmatrix}$$

or, expand along the third row:

$$\begin{vmatrix} a_{11} & a_{12} & a_{13} \\ a_{21} & a_{22} & a_{23} \\ a_{31} & a_{32} & a_{33} \end{vmatrix} = a_{31}\begin{vmatrix} a_{12} & a_{13} \\ a_{22} & a_{23} \end{vmatrix} - a_{32}\begin{vmatrix} a_{11} & a_{13} \\ a_{21} & a_{23} \end{vmatrix}$$

$$+ a_{33}\begin{vmatrix} a_{11} & a_{12} \\ a_{21} & a_{22} \end{vmatrix}$$

All of these give the same answer, as you should check by further evaluating the remaining $2 \times 2$ determinants!

---

Now that we understand how to evaluate determinants, we can state Cramer's rule for the solution of simultaneous equations in (3.1). For two equations, Cramer's rule gives the solution for the two unknowns, $x_1$ and $x_2$, as

$$x_1 = \frac{\begin{vmatrix} b_1 & a_{12} \\ b_2 & a_{22} \end{vmatrix}}{\begin{vmatrix} a_{11} & a_{12} \\ a_{21} & a_{22} \end{vmatrix}} = \frac{b_1 a_{22} - b_2 a_{12}}{a_{11} a_{22} - a_{12} a_{21}}$$

$$x_2 = \frac{\begin{vmatrix} a_{11} & b_1 \\ a_{21} & b_2 \end{vmatrix}}{\begin{vmatrix} a_{11} & a_{12} \\ a_{21} & a_{22} \end{vmatrix}} = \frac{b_2 a_{11} - b_1 a_{21}}{a_{11} a_{22} - a_{12} a_{21}}$$

(3.4)

In other words, Cramer's rule gives the solution for an unknown as the ratio of two determinants. The denominator determinant is simply the determinant of the coefficients of the equations in (3.1a). The numerator determinant is the determinant of the coefficients but with the right-hand side of the equations replacing the *column* of the unknown being solved for. For example, in solving for $x_1$, we replace the first column of the determinant of the coefficients in the numerator with $\begin{matrix} b_1 \\ b_2 \end{matrix}$. In solving for $x_2$, we replace the second column of the determinant of the coefficients in the numerator with $\begin{matrix} b_1 \\ b_2 \end{matrix}$.

---

*Example*

Solve the following two equations using Cramer's rule:

$$2x - 3y = 4$$

$$-x - 5y = 6$$

$$x = \frac{\begin{vmatrix} 4 & -3 \\ 6 & -5 \end{vmatrix}}{\begin{vmatrix} 2 & -3 \\ -1 & -5 \end{vmatrix}} = \frac{4 \times (-5) - (-3) \times 6}{2 \times (-5) - (-1) \times (-3)} = \frac{-2}{-13} = \frac{2}{13}$$

$$y = \frac{\begin{vmatrix} 2 & 4 \\ -1 & 6 \end{vmatrix}}{\begin{vmatrix} 2 & -3 \\ -1 & -5 \end{vmatrix}} = \frac{2 \times 6 - (-1) \times 4}{2 \times (-5) - (-1) \times (-3)} = \frac{16}{-13} = -\frac{16}{13}$$

You should "sanity check" these results.

---

For three equations, Cramer's rule is virtually identical to that for two equations. We form the solution for one of the unknowns as the ratio of two determinants. The denominator determinant is the determinant of the coefficients, and the numerator determinant is the determinant of the coefficients but with the right-hand side of the equations replacing the *column* of the unknown being solved for. For example, to solve the three equations in (3.1b), we form

$$
x_1 = \frac{\begin{vmatrix} b_1 & a_{12} & a_{13} \\ b_2 & a_{22} & a_{23} \\ b_3 & a_{32} & a_{33} \end{vmatrix}}{\begin{vmatrix} a_{11} & a_{12} & a_{13} \\ a_{21} & a_{22} & a_{23} \\ a_{31} & a_{32} & a_{33} \end{vmatrix}}
$$

$$
x_2 = \frac{\begin{vmatrix} a_{11} & b_1 & a_{13} \\ a_{21} & b_2 & a_{23} \\ a_{31} & b_3 & a_{33} \end{vmatrix}}{\begin{vmatrix} a_{11} & a_{12} & a_{13} \\ a_{21} & a_{22} & a_{23} \\ a_{31} & a_{32} & a_{33} \end{vmatrix}}
\tag{3.5}
$$

$$
x_3 = \frac{\begin{vmatrix} a_{11} & a_{12} & b_1 \\ a_{21} & a_{22} & b_2 \\ a_{31} & a_{32} & b_3 \end{vmatrix}}{\begin{vmatrix} a_{11} & a_{12} & a_{13} \\ a_{21} & a_{22} & a_{23} \\ a_{31} & a_{32} & a_{33} \end{vmatrix}}
$$

---

### *Example*

Solve the following three equations for $z$ by using Cramer's rule.

$$x + 3y - z = 4$$
$$2x - 4y + 2z = 3$$
$$3x + y + z = 2$$

Form

$$z = \frac{\begin{vmatrix} 1 & 3 & 4 \\ 2 & -4 & 3 \\ 3 & 1 & 2 \end{vmatrix}}{\begin{vmatrix} 1 & 3 & -1 \\ 2 & -4 & 2 \\ 3 & 1 & 1 \end{vmatrix}} = \frac{1\begin{vmatrix} -4 & 3 \\ 1 & 2 \end{vmatrix} - 3\begin{vmatrix} 2 & 3 \\ 3 & 2 \end{vmatrix} + 4\begin{vmatrix} 2 & -4 \\ 3 & 1 \end{vmatrix}}{1\begin{vmatrix} -4 & 2 \\ 1 & 1 \end{vmatrix} - 3\begin{vmatrix} 2 & 2 \\ 3 & 1 \end{vmatrix} - 1\begin{vmatrix} 2 & -4 \\ 3 & 1 \end{vmatrix}} = \frac{60}{-8} = -\frac{15}{2}$$

You should also solve for $x$ and $y$ and "sanity check" these results.

---

Cramer's rule is also very useful for solving *symbolic equations* where the known quantities, the $a$'s and the $b$'s, are symbols rather than numbers. This is particularly useful when using the Laplace transform to solve differential equations (Chapter 8). The result is given in terms of the Laplace transform variable $s$. Once we obtain the Laplace transform for the answer, we obtain the inverse transform of it (which is the desired answer).

---

### *Example*

Solve the Laplace-transformed equations

$$(2s+4)I_1(s) - 4I_2(s) = \frac{1}{s}$$

$$-4I_1(s) + (3s+5)I_2(s) = 0$$

$$I_1(s) = \frac{\begin{vmatrix} \dfrac{1}{s} & -4 \\ 0 & 3s+5 \end{vmatrix}}{\begin{vmatrix} 2s+4 & -4 \\ -4 & 3s+5 \end{vmatrix}}$$

$$= \frac{\dfrac{1}{s} \times (3s+5)}{(2s+4) \times (3s+5) - (-4) \times (-4)}$$

$$= \frac{3s+5}{s(6s^2 + 22s + 4)}$$

We could use substitution to achieve a solution, but Cramer's rule is a more systematic method of solution.

---

## 3.4  GAUSS ELIMINATION

Digital computers and calculators do not use Cramer's rule to solve simultaneous equations. Instead, they use the method of Gauss elimination, which is a much more efficient technique, particularly for more than three equations.

The method of *Gauss elimination* seeks to reduce the equations to *upper triangular form* using *elementary row operations*:

1. You can multiply an equation by any nonzero number.
2. You can add or subtract any two equations *and* replace one with that result.

These two operations do not change the solution to the equations. So we try to reduce, for example, the three equations in (3.1b) to *upper triangular* form:

$$\left.\begin{matrix} a_{11}x_1 + a_{12}x_2 + a_{13}x_3 = b_1 \\ a_{21}x_1 + a_{22}x_2 + a_{23}x_3 = b_2 \\ a_{31}x_1 + a_{32}x_2 + a_{33}x_3 = b_3 \end{matrix}\right\} \Rightarrow \left.\begin{matrix} c_{11}x_1 + c_{12}x_2 + c_{13}x_3 = d_1 \\ 0 + c_{22}x_2 + c_{23}x_3 = d_2 \\ 0 + 0 + c_{33}x_3 = d_3 \end{matrix}\right. \tag{3.6}$$

Once this upper triangular form is achieved, we easily solve for the unknowns using *back substitution*:

$$\begin{matrix} x_3 = \dfrac{d_3}{c_{33}} \\[2mm] x_2 = \dfrac{d_2}{c_{22}} - \dfrac{c_{23}}{c_{22}}x_3 \\[2mm] x_1 = \dfrac{d_1}{c_{11}} - \dfrac{c_{12}}{c_{11}}x_2 - \dfrac{c_{13}}{c_{11}}x_3 \end{matrix} \tag{3.7}$$

*Example*

Solve the following equations by Gauss elimination:

$$x + 3y - z = 4$$
$$2x - 4y + 2z = 3$$
$$3x + y + z = 2$$

Multiply row 1 by –2, add the result to row 2, and replace row 2 with the result:

$$x + 3y - z = 4$$
$$0 - 10y + 4z = -5$$
$$3x + y + z = 2$$

Now multiply row 1 by –3, add the result to row 3, and replace row 3 with the result:

$$x + 3y - z = 4$$
$$0 - 10y + 4z = -5$$
$$0 - 8y + 4z = -10$$

Now multiply row 2 by –8/10, add the result to row 3, and replace row 3 with the result:

$$x + 3y - z = 4$$
$$0 - 10y + 4z = -5$$
$$0 + 0 + \frac{4}{5}z = -6$$

Now solve by back substitution:

$$z = \frac{-6}{4/5} = -\frac{15}{2}$$

$$-10y + 4\left(-\frac{15}{2}\right) = -5 \quad \text{or} \quad y = -\frac{5}{2}$$

$$x + 3\left(-\frac{5}{2}\right) - \left(-\frac{15}{2}\right) = 4 \quad \text{or} \quad x = 4$$

The reader should "sanity check" these results.

---

### 3.5  MATRIX ALGEBRA

A *matrix* is an ordered *array* of elements, such as

$$\mathbf{A}_1 = \begin{bmatrix} a_{11} & a_{12} \\ a_{21} & a_{22} \end{bmatrix} \qquad \mathbf{A}_2 = \begin{bmatrix} a_{11} \\ a_{21} \\ a_{31} \end{bmatrix} \qquad \mathbf{A}_3 = \begin{bmatrix} a_{11} & a_{12} & a_{13} \\ a_{21} & a_{22} & a_{23} \end{bmatrix}$$

A matrix with $n$ rows and $m$ columns is said to have dimension $n \times m$. For the three matrices above, $\mathbf{A}_1$ is $2 \times 2$, $\mathbf{A}_2$ is $3 \times 1$, and $\mathbf{A}_3$ is $2 \times 3$. The entry in the $i$th row and $j$th column of a matrix is denoted by

$$[\mathbf{A}]_{ij} = a_{ij}$$

There a few special types of matrices. These are the $n \times n$ *square matrix* that has $n$ rows and $n$ columns:

$$\mathbf{A} = \begin{bmatrix} a_{11} & a_{12} & a_{13} \\ a_{21} & a_{22} & a_{23} \\ a_{31} & a_{32} & a_{33} \end{bmatrix}$$

the *symmetric matrix* ($a_{ij} = a_{ji}$), which must be a square matrix to be symmetric (about its main diagonal),

$$\mathbf{A} = \begin{bmatrix} a_{11} & a_{12} & a_{13} \\ a_{12} & a_{22} & a_{23} \\ a_{13} & a_{23} & a_{33} \end{bmatrix} \qquad a_{ij} = a_{ji}$$

and the *identity matrix*, which must be square:

$$\mathbf{1}_3 = \begin{bmatrix} 1 & 0 & 0 \\ 0 & 1 & 0 \\ 0 & 0 & 1 \end{bmatrix}$$

We now state the rules of *matrix algebra*.

***Addition***    To add two matrices, we simply add the corresponding elements. For

$$\mathbf{A} = \begin{bmatrix} a_{11} & a_{12} & a_{13} \\ a_{21} & a_{22} & a_{23} \end{bmatrix}$$

and

$$\mathbf{B} = \begin{bmatrix} b_{11} & b_{12} & b_{13} \\ b_{21} & b_{22} & b_{23} \end{bmatrix}$$

the sum is

$$\mathbf{A} + \mathbf{B} = \begin{bmatrix} a_{11} + b_{11} & a_{12} + b_{12} & a_{13} + b_{13} \\ a_{21} + b_{21} & a_{22} + b_{22} & a_{23} + b_{23} \end{bmatrix}$$

Note that in order to add two matrices, both must have the same dimensions: $n \times m$. The resulting matrix is of dimension of either (i.e., $n \times m + n \times m = n \times m$).

**Subtraction**    To subtract two matrices we simply subtract the corresponding elements:

$$\mathbf{A} - \mathbf{B} = \begin{bmatrix} a_{11} - b_{11} & a_{12} - b_{12} & a_{13} - b_{13} \\ a_{21} - b_{21} & a_{22} - b_{22} & a_{23} - b_{23} \end{bmatrix}$$

**Multiplication**    Multiplication of two matrices is a bit more complicated. Multiplication is best illustrated by the following diagram:

$$\mathbf{A} \times \mathbf{B} = \underbrace{\begin{bmatrix} \rightarrow \\ \rightarrow \\ \vdots \end{bmatrix}}_{\mathbf{A}} \times \underbrace{\begin{bmatrix} \downarrow & \downarrow & \cdots \end{bmatrix}}_{\mathbf{B}}$$

In other words, we take the *dot product* of each *row* of **A** and each *column* of **B**:

$$[\mathbf{A} \times \mathbf{B}]_{ij} = \sum_{k=1}^{n} a_{ik} b_{kj}$$

where the number of columns of **A** and the number of rows of **B** is $n$. For example, multiplying the two matrices

$$\mathbf{A} = \begin{bmatrix} a_{11} & a_{12} & a_{13} \\ a_{21} & a_{22} & a_{23} \end{bmatrix} \qquad \mathbf{B} = \begin{bmatrix} b_{11} & b_{12} \\ b_{21} & b_{22} \\ b_{31} & b_{32} \end{bmatrix}$$

*in the order* **A** × **B**, the result is

$$\mathbf{A} \times \mathbf{B} = \begin{bmatrix} a_{11}b_{11} + a_{12}b_{21} + a_{13}b_{31} & a_{11}b_{12} + a_{12}b_{22} + a_{13}b_{32} \\ a_{21}b_{11} + a_{22}b_{21} + a_{23}b_{31} & a_{21}b_{12} + a_{22}b_{22} + a_{23}b_{32} \end{bmatrix}$$

The order in which the two matrices are multiplied, **A** × **B** or **B** × **A**, is obviously important and does not give the same result: **A** × **B** ≠ **B** × **A**. In order to multiply two matrices as **A** × **B**, *the number of columns of* **A** *must equal the number of rows of* **B**. The dimension of the result is (*number of rows of* **A**) × (*number of columns of* **B**). For the two matrices above, **A** is $2 \times 3$ and **B** is $3 \times 2$, so that **A** × **B** is $2 \times 2$ but **B** × **A** is $3 \times 3$.

---

**Example**

Determine the product of the following two matrices as **A** × **B**:

$$A = \begin{bmatrix} 4 & -1 \\ 2 & 3 \\ -2 & 5 \end{bmatrix} \qquad B = \begin{bmatrix} 3 & 2 \\ -1 & -2 \end{bmatrix}$$

The product $A \times B$ is

$$A \times B = \begin{bmatrix} 4\times3+(-1)\times(-1) & 4\times2+(-1)\times(-2) \\ 2\times3+3\times(-1) & 2\times2+3\times(-2) \\ (-2)\times3+5\times(-1) & (-2)\times2+5\times(-2) \end{bmatrix} = \begin{bmatrix} 13 & 10 \\ 3 & -2 \\ -11 & -14 \end{bmatrix}$$

---

***Division (Matrix Inverse)***   The inverse of a scalar (a real number) is defined such that

$$a \times a^{-1} = a^{-1} \times a = 1$$

Similarly, the *inverse of a matrix* is defined such that

$$A \times A^{-1} = A^{-1} \times A = 1_n \qquad (3.8)$$

where the $A$ is $n \times n$ and $1_n$ is the $n \times n$ *identity matrix.*

How do we calculate $A^{-1}$ so that (3.8) is satisfied? A simple way to do this is to use Cramer's rule. Since $A^{-1}$ is $n \times n$, it has $n$ columns each of which we denote as an $n \times 1$ matrix (a vector) $C_i$. Then write the form of the inverse as

$$
\begin{aligned}
A^{-1} &= [C_1 \quad C_2 \quad \cdots \quad C_n] \\
&= \begin{bmatrix} c_{11} & c_{12} & \cdots & c_{1n} \\ c_{21} & c_{22} & \cdots & c_{2n} \\ \vdots & \vdots & \cdots & \vdots \\ c_{n1} & c_{n2} & \cdots & c_{nn} \end{bmatrix}
\end{aligned}
\qquad (3.9a)
$$

where we denote the *i*th column of $C$ as $C_i$:

$$C_i = \begin{bmatrix} c_{1i} \\ c_{2i} \\ \vdots \\ c_{ni} \end{bmatrix} \qquad (3.9b)$$

Now we solve for the columns of $C = A^{-1}$ by forming $A \times A^{-1} = A \times C = 1_n$ or

$$\mathbf{A} \times \mathbf{C} = [\mathbf{AC}_1 \quad \mathbf{AC}_2 \quad \cdots \quad \mathbf{AC}_n] = \begin{bmatrix} 1 & 0 & \cdots & 0 \\ 0 & 1 & \cdots & 0 \\ \vdots & \vdots & \ddots & \vdots \\ 0 & 0 & \cdots & 1 \end{bmatrix}$$

By matching columns on the right- and left-hand sides, we get $n$ equations that we can solve by Cramer's rule for the entries in each column $\mathbf{C}_i$:

$$\mathbf{A} \times \mathbf{C}_1 = \begin{bmatrix} 1 \\ 0 \\ \vdots \\ 0 \end{bmatrix}, \quad \mathbf{A} \times \mathbf{C}_2 = \begin{bmatrix} 0 \\ 1 \\ \vdots \\ 0 \end{bmatrix}, \quad \cdots \quad \mathbf{A} \times \mathbf{C}_n = \begin{bmatrix} 0 \\ 0 \\ \vdots \\ 1 \end{bmatrix} \qquad (3.10)$$

From this we can generate some rules for quickly obtaining the inverse of $\mathbf{A}, \mathbf{A}^{-1}$.

1. First obtain the *transpose* of $\mathbf{A}$, denoted as $\mathbf{A}'$, which we obtain by rotating $\mathbf{A}$ about its main diagonal $[\mathbf{A}']_{ij} = [\mathbf{A}]_{ji}$:

$$\mathbf{A}' = \begin{bmatrix} a_{11} & a_{21} & a_{31} \\ a_{12} & a_{22} & a_{32} \\ a_{13} & a_{23} & a_{33} \end{bmatrix} \qquad [\mathbf{A}']_{ij} = [\mathbf{A}]_{ji}$$

2. Then cross out the row and column of the entry of $\mathbf{A}^{-1}$ that is desired, and take the determinant of the remainder. For example:

$$[\mathbf{A}^{-1}]_{12} \Rightarrow \underbrace{\begin{bmatrix} X & X & X \\ a_{12} & X & a_{32} \\ a_{13} & X & a_{33} \end{bmatrix}}_{\mathbf{A}'}$$

$$= \begin{vmatrix} a_{12} & a_{32} \\ a_{13} & a_{33} \end{vmatrix}$$

$$= a_{12} \times a_{33} - a_{32} \times a_{13}$$

3. Multiply this result by $\pm$ according to the "checkerboard sign pattern":

$$\begin{vmatrix} + & - & + & \cdots \\ - & + & - & \cdots \\ + & - & + & \cdots \\ \vdots & \vdots & \vdots & \ddots \end{vmatrix}$$

to give, for example,

$$[\mathbf{A}^{-1}]_{12} \Rightarrow (-1)^{(1+2)} \begin{bmatrix} X & X & X \\ a_{12} & X & a_{32} \\ a_{13} & X & a_{33} \end{bmatrix}$$
$$\underbrace{\qquad\qquad}_{\mathbf{A}^t}$$

$$= (-1) \begin{vmatrix} a_{12} & a_{32} \\ a_{13} & a_{33} \end{vmatrix}$$

$$= -a_{12} \times a_{33} + a_{32} \times a_{13}$$

4. Divide the result by the determinant of **A**, |**A**|:

$$[\mathbf{A}^{-1}]_{12} = \frac{-a_{12} \times a_{33} + a_{32} \times a_{13}}{|\mathbf{A}|}$$

---

### *Example*

Determine the inverse of the following $3 \times 3$ matrix:

$$\mathbf{A} = \begin{bmatrix} 1 & 3 & -1 \\ 2 & -4 & 2 \\ 3 & 1 & 1 \end{bmatrix}$$

First, determine the determinant of the matrix:

$$|\mathbf{A}| = \begin{vmatrix} 1 & 3 & -1 \\ 2 & -4 & 2 \\ 3 & 1 & 1 \end{vmatrix} = 1 \times \begin{vmatrix} -4 & 2 \\ 1 & 1 \end{vmatrix} - 3 \times \begin{vmatrix} 2 & 2 \\ 3 & 1 \end{vmatrix} + (-1) \times \begin{vmatrix} 2 & -4 \\ 3 & 1 \end{vmatrix} = -8$$

Next, determine the transpose of the matrix:

$$\mathbf{A}^t = \begin{bmatrix} 1 & 2 & 3 \\ 3 & -4 & 1 \\ -1 & 2 & 1 \end{bmatrix}$$

Now determine the inverse:

$$\mathbf{A}^{-1} = \frac{1}{|\mathbf{A}|} \begin{bmatrix} \begin{vmatrix} -4 & 1 \\ 2 & 1 \end{vmatrix} & (-1)\begin{vmatrix} 3 & 1 \\ -1 & 1 \end{vmatrix} & \begin{vmatrix} 3 & -4 \\ -1 & 2 \end{vmatrix} \\ (-1)\begin{vmatrix} 2 & 3 \\ 2 & 1 \end{vmatrix} & \begin{vmatrix} 1 & 3 \\ -1 & 1 \end{vmatrix} & (-1)\begin{vmatrix} 1 & 2 \\ -1 & 2 \end{vmatrix} \\ \begin{vmatrix} 2 & 3 \\ -4 & 1 \end{vmatrix} & (-1)\begin{vmatrix} 1 & 3 \\ 3 & 1 \end{vmatrix} & \begin{vmatrix} 1 & 2 \\ 3 & -4 \end{vmatrix} \end{bmatrix}$$

$$= \frac{1}{|\mathbf{A}|} \begin{bmatrix} -6 & -4 & 2 \\ 4 & 4 & -4 \\ 14 & 8 & -10 \end{bmatrix}$$

$$= \begin{bmatrix} 3/4 & 1/2 & -1/4 \\ -1/2 & -1/2 & 1/2 \\ -7/4 & -1 & 5/4 \end{bmatrix}$$

The reader should "sanity check" this result by showing that $\mathbf{A} \times \mathbf{A}^{-1} = \mathbf{A}^{-1} \times \mathbf{A} = \mathbf{1}_3$.

---

The matrix inverse can be used to solve simultaneous algebraic equations. To do so we write the equations (e.g., three equations):

$$a_{11}x_1 + a_{12}x_2 + a_{13}x_3 = b_1$$
$$a_{21}x_1 + a_{22}x_2 + a_{23}x_3 = b_2$$
$$a_{31}x_1 + a_{32}x_2 + a_{33}x_3 = b_3$$

in matrix form as

$$\boxed{\mathbf{AX} = \mathbf{B}} \tag{3.11a}$$

where

$$\mathbf{A} = \begin{bmatrix} a_{11} & a_{12} & a_{13} \\ a_{21} & a_{22} & a_{23} \\ a_{31} & a_{32} & a_{33} \end{bmatrix} \tag{3.11b}$$

$$\mathbf{X} = \begin{bmatrix} x_1 \\ x_2 \\ x_3 \end{bmatrix} \tag{3.11c}$$

$$\mathbf{B} = \begin{bmatrix} b_1 \\ b_2 \\ b_3 \end{bmatrix} \tag{3.11d}$$

(The reader should "sanity check" this.) To obtain the solution, *premultiply* (multiply on the left) both sides by $\mathbf{A}^{-1}$ to give

$$\mathbf{A}^{-1} \times \mathbf{A}\mathbf{X} = \mathbf{A}^{-1} \times \mathbf{B}$$

This gives

$$\boxed{\mathbf{X} = \mathbf{A}^{-1}\mathbf{B}} \tag{3.12}$$

since

$$\mathbf{A}^{-1} \times \mathbf{A} = \mathbf{1}_n$$

and

$$\mathbf{1}_n \mathbf{X} = \mathbf{X}$$

---

### *Example*

Solve, by using the matrix inverse, the equations

$$x + 3y - z = 4$$
$$2x - 4y + 2z = 3$$
$$3x + y + z = 2$$

These can be written in matrix form as $\mathbf{AX} = \mathbf{B}$:

$$\underbrace{\begin{bmatrix} 1 & 3 & -1 \\ 2 & -4 & 2 \\ 3 & 1 & 1 \end{bmatrix}}_{\mathbf{A}} \underbrace{\begin{bmatrix} x \\ y \\ z \end{bmatrix}}_{\mathbf{X}} = \underbrace{\begin{bmatrix} 4 \\ 3 \\ 2 \end{bmatrix}}_{\mathbf{B}}$$

Premultiplying both sides by $\mathbf{A}^{-1}$ (which was obtained in the preceding example) gives the solution

$$\underbrace{\begin{bmatrix} x \\ y \\ z \end{bmatrix}}_{\mathbf{X}} = \underbrace{\begin{bmatrix} 3/4 & 1/2 & -1/4 \\ -1/2 & -1/2 & 1/2 \\ -7/4 & -1 & 5/4 \end{bmatrix}}_{\mathbf{A}^{-1}} \times \underbrace{\begin{bmatrix} 4 \\ 3 \\ 2 \end{bmatrix}}_{\mathbf{B}}$$

$$= \begin{bmatrix} 4 \\ -5/2 \\ -15/2 \end{bmatrix}$$

The reader should "sanity check" this solution by substituting into the original equations.

---

The matrix inverse is particularly simple to obtain for the special case of a $2 \times 2$ matrix:

$$\mathbf{A} = \begin{bmatrix} a_{11} & a_{12} \\ a_{21} & a_{22} \end{bmatrix}$$

The matrix inverse is

$$\boxed{\mathbf{A}^{-1} = \frac{1}{|\mathbf{A}|} \begin{bmatrix} a_{22} & -a_{12} \\ -a_{21} & a_{11} \end{bmatrix}} \qquad (3.13)$$

Observe that the inverse of a $2 \times 2$ matrix is obtained by

1. swapping the main-diagonal terms
2. negating the off-diagonal terms
3. dividing all terms by the determinant of $\mathbf{A}$

In matrix form the two equations are

$$\underbrace{\begin{bmatrix} a_{11} & a_{12} \\ a_{21} & a_{22} \end{bmatrix}}_{\mathbf{A}} \underbrace{\begin{bmatrix} x_1 \\ x_2 \end{bmatrix}}_{\mathbf{X}} = \underbrace{\begin{bmatrix} b_1 \\ b_2 \end{bmatrix}}_{\mathbf{B}}$$

Premultiplying both sides by $\mathbf{A}^{-1}$ gives the solution as

$$\mathbf{X} = \mathbf{A}^{-1}\mathbf{B}$$

or

$$\begin{bmatrix} x_1 \\ x_2 \end{bmatrix} = \frac{1}{|\mathbf{A}|} \begin{bmatrix} a_{22} & -a_{12} \\ -a_{21} & a_{11} \end{bmatrix} \begin{bmatrix} b_1 \\ b_2 \end{bmatrix}$$

$$= \frac{1}{a_{11}a_{22} - a_{12}a_{21}} \begin{bmatrix} a_{22}b_1 - a_{12}b_2 \\ -a_{21}b_1 + a_{11}b_2 \end{bmatrix}$$

(3.14)

whose solutions (as were obtained with Cramer's rule) are

$$x_1 = \frac{a_{22}b_1 - a_{12}b_2}{a_{11}a_{22} - a_{12}a_{21}}$$

(3.15a)

$$x_2 = \frac{-a_{21}b_1 + a_{11}b_2}{a_{11}a_{22} - a_{12}a_{21}}$$

(3.15b)

***Example***

Determine the solution to the following two equations by matrix inverse:

$$2x - 3y = 4$$
$$-x - 5y = 6$$

Writing these in matrix form gives

$$\underbrace{\begin{bmatrix} 2 & -3 \\ -1 & -5 \end{bmatrix}}_{\mathbf{A}} \underbrace{\begin{bmatrix} x \\ y \end{bmatrix}}_{\mathbf{X}} = \underbrace{\begin{bmatrix} 4 \\ 6 \end{bmatrix}}_{\mathbf{B}}$$

Premultiplying by $\mathbf{A}^{-1}$ gives the solution as

$$\begin{bmatrix} x \\ y \end{bmatrix} = \frac{1}{|\mathbf{A}|} \begin{bmatrix} -5 & 3 \\ 1 & 2 \end{bmatrix} \begin{bmatrix} 4 \\ 6 \end{bmatrix}$$

$$= \frac{1}{2 \times (-5) - (-1) \times (-3)} \begin{bmatrix} (-5) \times 4 + 3 \times 6 \\ 1 \times 4 + 2 \times 6 \end{bmatrix}$$

$$= \frac{1}{-13} \begin{bmatrix} -2 \\ 16 \end{bmatrix}$$

$$= \begin{bmatrix} 2/13 \\ -16/13 \end{bmatrix}$$

The reader should "sanity check" this result.

Obtaining the solution to simultaneous algebraic equations by either Cramer's rule or the matrix inverse, $\mathbf{A}^{-1}$, method involves the determinant of the coefficient matrix, $|\mathbf{A}|$, which appears in the denominators of the solutions. It may happen that $|\mathbf{A}| = 0$. In this case the coefficient matrix, $\mathbf{A}$, is said to be *singular*. What does this mean with regard to the solution of those equations? To interpret this result, consider the case of two equations. At the beginning of this chapter we interpreted the solution of two equations as the intersection of two straight lines in two-dimensional space, which is what the two equations represent (see Figure 3.1). (Three equations would represent three planes in three-dimensional space.) For the case of two equations, they have either (1) unique solutions when the two lines intersect at only one point (Figure 3.1), or (2) no solutions when the two lines are parallel and do not intersect (Figure 3.2), or (3) an infinite number of solutions when the two lines are identical (Figure 3.2). If $\mathbf{A}$ is nonsingular, $|\mathbf{A}| \neq 0$, we clearly will have the unique solution: one and only one intersection of the two lines. On the other hand, if $|\mathbf{A}| = 0$, this must mean that the two lines are parallel so that there are either no solutions (no intersections of the two lines) or an infinite number of solutions (an infinite number of intersections of the two lines). For example, suppose that one equation of the set is a multiple of the other equation:

$$a_{11}x_1 + a_{12}x_2 = b_1$$
$$ka_{11}x_1 + ka_{12}x_2 = kb_1$$

In this case the two lines are parallel and lie on top of each other, resulting in an infinite number of solutions. For this case the determinant of the coefficients is zero:

$$|\mathbf{A}| = \begin{vmatrix} a_{11} & a_{12} \\ ka_{11} & ka_{12} \end{vmatrix} = 0$$

On the other hand, suppose that the right-hand side in the second equation above is $b_2 \neq kb_1$. For this case the coefficient matrix is again singular, but this represents two lines that are parallel but do not lie on top of each other. Hence, there are no solutions. You can see that this is a general result for $n$ linear algebraic equations: If $|\mathbf{A}| \neq 0$, there are $n$ unique solutions for the $x_i$, but for $|\mathbf{A}| = 0$ there are either no solutions or an infinite number of solutions for the $x_i$.

There is one final implication of $|\mathbf{A}| = 0$. Suppose that the right-hand sides of the two equations are zero: $b_1 = b_2 = 0$. In this case, the two lines that the equations represent pass through the origin of the coordinate system so that they intersect at $x_1 = x_2 = 0$. If $|\mathbf{A}| \neq 0$, they are not parallel, so that $x_1 = x_2 = 0$

is the only, unique solution. If $|A| = 0$, the two lines are parallel and lie on top of each other, giving an infinite number of solutions.

Finally, there are several important properties of matrix algebra that we will enumerate. These allow the manipulation of matrices in the same fashion as scalar algebra. First, the product of matrices is distributive:

$$A(B+C) = AB + AC \qquad (3.16a)$$

and associative:

$$(AB)C = A(BC) \qquad (3.16b)$$

and the order of multiplication is obviously important:

$$AB \neq BA \qquad (3.16c)$$

In taking the products above, it is assumed, of course, that the dimensions of the matrices allow the multiplications indicated. For example, in taking the product $AB$, it is assumed that the number of columns of $A$ is the same as the number of rows of $B$.

Care must be exercised in extending results for scalars to the case of matrices. If $AB = 0$, it is not necessarily true that either $A$ or $B$ are zero matrices, and if $AC = AD$, it is not necessarily true that $C$ and $D$ are equal. For example,

$$A = [1 \quad 2] \qquad B = \begin{bmatrix} 4 \\ -2 \end{bmatrix} \qquad C = \begin{bmatrix} 3 \\ 1 \end{bmatrix} \qquad D = \begin{bmatrix} -3 \\ 4 \end{bmatrix}$$

For these matrices, $AB = 0$ and $AC = AD$, but in the first case, $B \neq 0$, and in the second case, $C \neq D$. However, suppose that $A, B, C,$ and $D$ are square, $n \times n$, and $A$ is nonsingular, $|A| \neq 0$, so that $A^{-1}$ exists. We can premultiply both sides of $AB = 0$ and $AC = AD$ by $A^{-1}$, showing that in this case, $B = 0$ and $C = D$, respectively.

The transpose of the transpose of a matrix gives back the original matrix: $(A')' = A$, and the inverse of the inverse of a matrix gives back the original matrix: $(A^{-1})^{-1} = A$. The transpose of the product of two matrices equals the transpose of the two matrices *in the reverse order:*

$$(AB)' = B'A' \qquad (3.16d)$$

as is the case for the inverse of a product:

$$(AB)^{-1} = B^{-1}A^{-1} \qquad (3.16e)$$

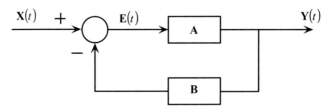

**Fig. 3.4.** An automatic control feedback positioning system.

It is very important to note that the inverse of the sum of two matrices *does not equal the sum of their inverses:*

$$(\mathbf{A} + \mathbf{B})^{-1} \neq \mathbf{A}^{-1} + \mathbf{B}^{-1} \tag{3.16f}$$

Matrices are very useful for manipulating large amounts of data and relationships in a very handy and compact manner. For example, consider the automatic control system shown in Figure 3.4. A set of $n$ inputs from a controller that are the entries in the $n \times 1$ vector $\mathbf{X}(t)$ are to direct the positions of a robotic arm and its fingers to precise positions in an automotive assembly line. These $n$ inputs are to be processed by a controller denoted as the block $\mathbf{A}$, which is $n \times n$. The outputs of $\mathbf{A}$ are the desired commands to the arm, which are denoted as the entries in the $n \times 1$ vector $\mathbf{Y}(t)$. The output $\mathbf{Y}(t)$ is fed back and processed by a $n \times n$ controller, denoted as $\mathbf{B}$, where it is subtracted from the input to give $n$ "error" signals, which are the $n$ entries in the $n \times 1$ vector $\mathbf{E}(t)$. The block $\mathbf{A}$ is known, and the design problem is to determine $\mathbf{B}$ to move the robot arm to the required positions in a very precise and rapid manner with minimal error. To investigate the system, we write the governing (matrix) equations for the system:

$$\mathbf{E}(t) = \mathbf{X}(t) - \mathbf{B}\mathbf{Y}(t)$$

and

$$\mathbf{Y}(t) = \mathbf{A}\mathbf{E}(t)$$

Combining these gives the relation between $\mathbf{Y}$ and $\mathbf{X}$ (which you should verify):

$$\mathbf{Y}(t) = [\mathbf{1}_n + \mathbf{A}\mathbf{B}]^{-1}\mathbf{A}\mathbf{X}(t)$$

Hence, to adjust the position precisely, the controller (computer) **B** must respond in real time to satisfy this last result. Observe in this last result that **A** and **B** must be multiplied to give **AB**, which is then added to the $n \times n$ identity matrix $\mathbf{1}_n$, and the result is inverted. This then premultiplies **A** to give the overall system controller. Observe that $[\mathbf{1}_n + \mathbf{AB}]^{-1} \neq \mathbf{1}_n + [\mathbf{AB}]^{-1} \neq \mathbf{1}_n + \mathbf{A}^{-1}\mathbf{B}^{-1}$.

# 4 Solution of Linear, Constant-Coefficient, Ordinary Differential Equations

Engineering systems have inputs that are usually functions of time, denoted as $t$. *Static* engineering systems are generally governed by *algebraic equations.* *Dynamic* engineering systems are generally governed by *differential equations.* These differential equations contain derivatives with respect to time $t$ and are somewhat more difficult to solve than are algebraic equations. In this chapter we examine the most common type of differential equation that is frequently found in engineering systems: the *linear, constant-coefficient, ordinary differential equation.*

## 4.1 HOW TO IDENTIFY LINEAR, CONSTANT-COEFFICIENT, ORDINARY DIFFERENTIAL EQUATIONS

What do they look like? In these equations there is only one independent variable, which we denote as $t$ (perhaps symbolizing time). The function to be solved for is denoted as $x(t)$. Since the unknown is a function of $t$, any derivatives will be with respect to $t$: $dx(t)/dt$, $d^2x(t)/dt^2$, and so on. These are said to be *ordinary derivatives*, as opposed to partial derivatives. A derivative $d^nx(t)/dt^n$ is said to be of *order n.*

The general form of a *first-order* ordinary differential equation is

$$\frac{dx(t)}{dt} + ax(t) = f(t) \tag{4.1a}$$

The general form of a *second-order* ordinary differential equation is

$$\frac{d^2x(t)}{dt^2} + a\frac{dx(t)}{dt} + bx(t) = f(t) \tag{4.1b}$$

*Essential Math Skills for Engineers,* By Clayton R. Paul
Copyright © 2009 John Wiley & Sons, Inc.

The coefficients *a* and *b* are *constants* and are known (given). The right-hand-side function of $t$, $f(t)$, is also known (given). The right-hand side, $f(t)$, is called *the forcing function* since it represents the *source* that is driving the engineering system. The task will be to solve for a function $x(t)$ which when substituted into the equation, satisfies it. It is always a good idea to *make the leading coefficient unity*! All of our results apply, of course, in identical fashion to differential equations where the independent and dependent variables might have different symbols and meanings, such as $y(x)$. You should be alert to this throughout your engineering studies. It is embarrassing not to know how to solve the differential equation

$$\frac{dy(x)}{dx} + ay(x) = f(x)$$

once you have learned to solve the differential equation

$$\frac{dx(t)}{dt} + ax(t) = f(t)$$

as there are no differences in their solutions. We will consider only first- and second-order differential equations, meaning that the highest-order derivative will be first or second, respectively. Although higher-order differential equations occur frequently in engineering systems, they are only slightly more tedious to solve. However, you can easily extend to higher-order differential equations the methods we develop for first- and second-order equations.

The equation is said to be *linear* if neither $x(t)$ nor its derivatives are raised to a power greater than 1, and no products of $x(t)$ and its derivatives appear. Note that there is a *significant difference* between $d^2x(t)/dt^2$ and $[dx(t)/dt]^2$: The first is a second-order derivative and the second is a first-order derivative that is *squared*. An example of a *nonlinear* differential equation is

$$\frac{d^2x(t)}{dt^2} + a\underbrace{\left(\frac{dx(t)}{dt}\right)^2}_{\text{nonlinear}} + b\underbrace{x^3(t)}_{\text{nonlinear}} + c\underbrace{x(t)\frac{dx(t)}{dt}}_{\text{nonlinear}} = f(t)$$

The equation is *constant-coefficient* if *a* and *b* are *constants* that are not functions of *t*. An example of a *non-constant-coefficient* differential equation is

$$\frac{dx(t)}{dt} + (2t)x(t) = f(t)$$

If a differential equation is either (a) *non-constant-coefficient* and/or (b) *nonlinear*, it is extremely difficult to solve, and generally only numerical, approximate solutions are obtainable! As was the case with algebraic equations, nonlinear and/or time-varying (nonconstant coefficient) differential equations occur throughout all of the engineering disciplines. But because they are so difficult to solve that an attempt at hand solution yields little or no insight into how a system behaves, we generally resort to solving them numerically in an approximate fashion using a computer or a calculator. Numerical solutions of differential equations are discussed at the end of this chapter. Throughout all of this you should keep in mind that the primary goal is *to use mathematical skills to determine what the equations governing the system are telling us about how the system behaves.*

## 4.2   WHERE THEY ARISE: THE MEANING OF A SOLUTION

In Chapter 1 we showed examples of where differential equations arise in electrical and mechanical engineering systems. Ordinary differential equations can be found routinely in all of engineering.

How do we know that a particular formula for $x(t)$ is a solution to the differential equation? The simple method is to substitute it into the differential equation, perform the required differentiations, and see if satisfies the equation [i.e., both sides of (4.1) are the same]. However, unlike algebraic equations, differential equations have an infinite number of possible solutions. So we need some additional information to pin down which of these is the actual solution for the problem we are investigating. This additional information comprises what are called the *initial conditions* for the specific problem being investigated. For example, consider the first-order equation in (4.1a). For the specific problem that this differential equation describes, we would need to specify an additional initial condition for the value of $x(t)$ at some starting time, say $t = 0$, which we denote as $x(0)$. So the complete specification of the problem would be to specify (a) the differential equation *and* (b) the initial condition as

$$\boxed{\frac{dx(t)}{dt} + ax(t) = f(t) \qquad x(0)} \qquad (4.1a)$$

Once we have obtained a solution $x(t)$ that satisfies this differential equation *and* the initial condition, we can say that the solution $x(t)$ is valid for all $t \geq 0$. Similarly, the second-order equation in (4.1b) requires *two initial conditions:* $x(0)$ and its derivative at $t = 0, dx(t)/dt|_{t=0} \equiv \dot{x}(0)$. So the complete specification of a second-order equation would be

$$\boxed{\frac{d^2x(t)}{dt^2} + a\frac{dx(t)}{dt} + bx(t) = f(t) \qquad x(0), \quad \dot{x}(0)} \qquad (4.1b)$$

For $x(t)$ to be *the solution* of this, (1) it must satisfy the differential equation when it is substituted into the differential equation, and (2) $x(t)$ and its derivative must also satisfy the two initial conditions.

## 4.3  SOLUTION OF FIRST-ORDER EQUATIONS

The complete solution to a first-order equation that also satisfies the initial condition

$$\boxed{\frac{dx(t)}{dt} + ax(t) = f(t) \qquad x(0)} \tag{4.1a}$$

is the sum of a *homogeneous* (or *natural*) solution $x_h(t)$ that satisfies the equation with the right-hand side zero, $f(t) = 0$:

$$\boxed{\frac{dx_h(t)}{dt} + ax_h(t) = 0} \tag{4.2a}$$

and the *forced* (or *particular*) solution $x_f(t)$ that satisfies the equation with the right-hand side not zero, $f(t) \neq 0$:

$$\boxed{\frac{dx_f(t)}{dt} + ax_f(t) = f(t)} \tag{4.2b}$$

The *total solution* is

$$\boxed{x(t) = x_h(t) + x_f(t)} \tag{4.2c}$$

How can we prove this? Substitute (4.2c) into (4.1a) to give

$$\frac{d}{dt}[x_h(t) + x_f(t)] + a[x_h(t) + x_f(t)] \overset{?}{=} f(t)$$

Expand this as

$$\left[\frac{dx_h(t)}{dt} + ax_h(t)\right] + \left[\frac{dx_f(t)}{dt} + ax_f(t)\right] \overset{?}{=} [0] + [f(t)]$$

But this is the sum of two "truths," (4.2a) and (4.2b), which is therefore true.

This "superposition" of the solutions to the homogeneous and forced solutions does not work for a nonlinear differential equation. To show why it does not work for nonlinear differential equations, consider the nonlinear differential equation

$$\frac{dx}{dt} + ax^2 = f$$

Substituting the sum of the homogeneous and forced solutions gives

$$\frac{d}{dt}[x_h + x_f] + a[x_h + x_f]^2 \overset{?}{=} f$$

Squaring the nonlinear term gives

$$\frac{d}{dt}[x_h + x_f] + a[x_h^2 + 2x_h x_f + x_f^2] \overset{?}{=} f(t)$$

Expanding this gives

$$\left[\frac{dx_h}{dt} + ax_h^2\right] + \left[\frac{dx_f}{dt} + ax_f^2\right] + 2ax_h x_f \neq [0] + [f]$$

But this is not satisfied because there is an additional "cross product" term, $2ax_h x_f$, introduced by the nonlinear term of the differential equation that is not accounted for.

### 4.3.1   The Homogeneous Solution

The homogeneous solution satisfies (4.2a):

$$\boxed{\frac{dx_h(t)}{dt} + ax_h(t) = 0} \tag{4.2a}$$

What do we mean by a solution? Again, the simple answer is any function $x_h(t)$ which when substituted into (4.2a) satisfies the equation. Clearly, any such function must have the important property that *its derivative must give back the same form of function*. The only such form is the exponential

$$\boxed{x_h(t) = Ce^{pt}} \tag{4.3}$$

Substituting (4.3) into (4.2a) yields

$$pCe^{pt} + aCe^{pt} = (p+a)Ce^{pt} = 0$$

We cannot choose $C$ to be zero; otherwise, we end up with a trivial solution. The exponential $e^{pt}$ is never zero (truly zero) except at $t = \pm\infty$. So the only way that this can be satisfied is to satisfy the *characteristic equation*

$$p + a = 0$$

giving the value of $p$ as

$$p = -a \tag{4.4}$$

Hence, the form of the *homogeneous solution* is

$$\boxed{x_h(t) = Ce^{-at}} \tag{4.5}$$

The constant $C$ is, as yet, undetermined but will be evaluated by applying the initial condition $x(0)$ to the *total solution* in (4.2c) once we have determined the forced solution, which we do next.

### 4.3.2   The Forced Solution for "Nice" $f(t)$

Our task is now to determine the function $x_f(t)$ that satisfies (4.2b):

$$\boxed{\frac{dx_f(t)}{dt} + ax_f(t) = f(t)} \tag{4.2b}$$

In engineering problems we most frequently encounter what are called "nice" $f(t)$. These are (1) $f(t)$ is a constant as $f(t) = K$ where $K$ is a constant; (2) $f(t)$ is a polynomial in powers of $t$ as $f(t) = c_0 + c_1 t + c_2 t^2 + c_3 t^3 + \ldots + c_n t^n$; and (3) $f(t)$ is a sinusoid (sine or cosine) as $f(t) = A\sin(\omega t + \theta)$ or $f(t) = A\cos(\omega t + \theta)$. The simplest method for determining the forced solution for these three very common forcing functions is the "method of undetermined coefficients," where we assume a *form* for $x_f(t)$ that is the same *form* as $f(t)$ but with some undetermined constants in it. We substitute these forms into (4.2b), compare both sides and determine those undetermined constants.

*f(t) Is a Constant*   First determine the forced solution for a constant forcing function:

$$\boxed{f(t) = K} \tag{4.6a}$$

where $K$ is known (given). We assume the forced solution to be of the same form:

$$\boxed{x_f(t) = M} \tag{4.6b}$$

and substitute into (4.2b) to give $[dx_f(t)/dt = 0$ for this $f(t)]$:

$$aM = K$$

or

$$x_f(t) = \frac{K}{a} \tag{4.7}$$

*f(t) Is a Polynomial in t*   Next we determine the forced solution for a forcing function that is a polynomial in $t$:

$$f(t) = c_0 + c_1 t + c_2 t^2 + c_3 t^3 + \cdots + c_n t^n \tag{4.8a}$$

where the $c_0, c_1, c_2, \ldots, c_n$ are known (given). By the method of undetermined coefficients, we assume a form of the forced solution as

$$x_f(t) = d_0 + d_1 t + d_2 t^2 + d_3 t^3 + \cdots + d_n t^n \tag{4.8b}$$

Substituting (4.8b) into (4.2b) gives

$$(d_1 + 2d_2 t + 3d_3 t^2 + \cdots + n d_n t^{n-1}) + a(d_0 + d_1 t + d_2 t^2 + \cdots + d_n t^n)$$
$$= c_0 + c_1 t + c_2 t^2 + \cdots + c_n t^n$$

Matching the coefficients of corresponding powers of $t$ on both sides gives simultaneous algebraic equations to solve for $d_0, d_1, d_2, \ldots, d_n$.

***f(t) Is a Sinusoid***   Finally, we determine the forced solution for a forcing function that is a sinusoid:

$$f(t) = A\sin(\omega t + \theta) \quad \text{or} \quad f(t) = A\cos(\omega t + \theta) \tag{4.9}$$

By the method of undetermined coefficients we assume a solution for $x_f(t)$ which is also a sinusoid. We could assume a form of the forced solution as $x_f(t) = M\sin(\omega t + \phi)$ or $x_f(t) = M\cos(\omega t + \phi)$, substitute into (4.2b), and determine $M$ and $\phi$, but that would be "messy." We demonstrated in Section 2.8.1 a simpler method of obtaining this using complex algebra. Replace $f(t)$ with $f(t) = A\angle\theta e^{j\omega t}$ and solve

$$\frac{dX_f(t)}{dt} + aX_f(t) = A\angle\theta e^{j\omega t} \tag{4.10a}$$

Assume, by the method of undetermined coefficients, the forced solution to this equation to be

$$X_f(t) = M\angle\phi e^{j\omega t} \tag{4.10b}$$

Substitute (4.10b) into (4.10a) and determine the unknowns $M$ and $\phi$. Substituting (4.10b) into (4.10a) gives $\left[\text{recall that } \dfrac{d}{dt} e^{j\omega t} = (j\omega) e^{j\omega t}\right]$

$$j\omega M\angle\phi e^{j\omega t} + aM\angle\phi e^{j\omega t} = A\angle\theta e^{j\omega t}$$

Canceling the common $e^{j\omega t}$ and collecting terms gives

$$M\angle\phi = \frac{A\angle\theta}{j\omega + a} \tag{4.10c}$$

from which we obtain, using complex algebra, $M$ and $\phi$. Depending on whether $f(t)$ is a sine or a cosine, we then write

$$\boxed{x_f(t) = M \sin(\omega t + \phi) \quad \text{or} \quad x_f(t) = M \cos(\omega t + \phi)} \tag{4.11}$$

### 4.3.3    The Total Solution

Now we write the total solution as the sum of the homogeneous and forced solutions:

$$\begin{aligned} x(t) &= x_h(t) + x_f(t) \\ &= Ce^{-at} + x_f(t) \end{aligned} \tag{4.12}$$

Applying the initial condition to this total solution gives

$$x(0) = C + x_f(0)$$

which we solve for $C$ to give

$$C = x(0) - x_f(0)$$

The final solution valid for $t \geq 0$ is

$$\boxed{x(t) = [x(0) - x_f(0)]e^{-at} + x_f(t) \qquad \text{for } t \geq 0} \tag{4.13}$$

From an engineering perspective, it is very important to visualize and understand what the total solution in (4.13) is telling us about the behavior of the solution. Note that the total solution consists of the sum of two parts. The first part is the homogeneous solution and varies with $t$ in an exponential fashion as $e^{-at}$. If $a$ is a positive number, this exponential part of the solution goes to zero as time increases: $t \to \infty$. It goes to zero exponentially, leaving the forced solution $x(t) \to x_f(t)$ as $t \to \infty$. For example, for the differential equation

$$\frac{dx(t)}{dt} + 2x(t) = 30 \qquad x(0) = 3$$

the forced solution is $x_f(t) = 15$ and the total solution is

$$x(t) = \underbrace{-12e^{-2t}}_{x_h(t)} + \underbrace{15}_{x_f(t)} \qquad \text{for } t \geq 0$$

You should "sanity check" this by substituting it into the differential equation and also checking that it satisfies the initial condition. So the solution should start at 3 at $t = 0$ and should increase exponentially to the forced solution of 15 as illustrated in Figure 4.1.

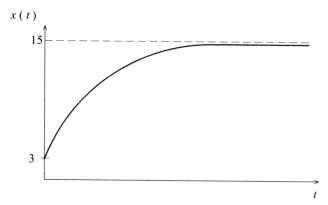

**Fig. 4.1.** A plot of the solution $x(t) = -12e^{-2t} + 15$ for $x(0) = 3$.

### Example

Determine the solution to the following differential equation having the initial condition $x(0) = 4$:

$$\frac{dx(t)}{dt} + \underset{a}{2}\,x(t) = \underset{f(t)}{10} \qquad x(0) = 4$$

The homogeneous solution is

$$x_h(t) = Ce^{-2t}$$

Assume the form of the forced solution to be

$$x_f(t) = M$$

and substitute to give

$$0 + 2M = 10$$

The forced solution is

$$x_f(t) = M = 5$$

The total solution is

$$x(t) = x_h(t) + x_f(t)$$
$$= Ce^{-2t} + 5$$

Applying the initial condition gives

$$4 = Ce^{-2(0)} + 5$$

so that $C = -1$, and hence the total solution valid for $t \geq 0$ is

$$x(t) = -e^{-2t} + 5 \qquad \text{for} \quad t \geq 0$$

The reader should "sanity check" this result.

---

### *Example*

Determine the solution to the following differential equation having the initial condition $x(0) = 2$:

$$\frac{dx(t)}{dt} + \underbrace{3x(t)}_{a} = \underbrace{5t}_{f(t)} \qquad x(0) = 2$$

The homogeneous solution is

$$x_h(t) = Ce^{-3t}$$

Assume the form of the forced solution to be

$$x_f(t) = M_1 + M_2 t$$

[Note that we must use a form that also contains all lower powers of $t$ if $f(t)$ contains a power of $t$.] Substitute to give

$$M_2 + 3M_1 + 3M_2 t = 5t$$

Matching coefficients of powers of $t$ gives two simultaneous algebraic equations to be solved for $M_1$ and $M_2$:

$$3M_1 + M_2 = 0$$

$$3M_2 = 5$$

which when solved give $M_1 = -5/9$ and $M_2 = 5/3$. Hence, the forced solution is

$$x_f(t) = -\frac{5}{9} + \frac{5}{3}t$$

The total solution is

$$x(t) = x_h(t) + x_f(t)$$

$$= Ce^{-3t} - \frac{5}{9} + \frac{5}{3}t$$

Applying the initial condition gives

$$2 = Ce^{-3(0)} - \frac{5}{9}$$

so that $C = 23/9$ and hence the total solution valid for $t \geq 0$ is

$$x(t) = \frac{23}{9}e^{-3t} - \frac{5}{9} + \frac{5}{3}t \qquad \text{for} \quad t \geq 0$$

The reader should "sanity check" this result.

---

### *Example*

Determine the solution to the following differential equation having the initial condition $x(0) = 1$:

$$\underbrace{\frac{dx(t)}{dt} + \underset{a}{5}x(t)} = \underbrace{10\cos 2t}_{f(t)} \qquad x(0) = 1$$

The homogeneous solution is

$$x_h(t) = Ce^{-5t}$$

To obtain the forced solution, we replace $f(t)$ with

$$f(t) \Rightarrow 10e^{j2t}$$

and solve, instead,

$$\frac{d\mathbf{X}_f(t)}{dt} + 5\mathbf{X}_f(t) = 10e^{j2t}$$

We assume the forced solution to be of the form

$$\mathbf{X}_f(t) = M\angle\phi e^{j2t}$$

Substitute to give

$$M\angle\phi = \frac{10}{j2+5}$$
$$= 1.86\angle -21.8°$$

Hence, the forced solution is [since $f(t)$ was a cosine]

$$x_f(t) = 1.86\cos(2t - 21.8°)$$

The total solution is

$$x(t) = x_h(t) + x_f(t)$$
$$= Ce^{-5t} + 1.86\cos(2t - 21.8°)$$

Applying the initial condition gives

$$1 = Ce^{-5(0)} + 1.86\cos(0 - 21.8°)$$

so that $C = -0.724$, and hence the total solution valid for $t \geq 0$ is

$$x(t) = -0.724e^{-5t} + 1.86\cos(2t - 21.8°) \qquad \text{for} \quad t \geq 0$$

The reader should "sanity check" this result.

---

### 4.3.4  A Special Case

> If $f(t)$ has the same *form* as the homogeneous solution, $f(t) = Ke^{-at}$, the usual assumed form of the forced solution must be multiplied by $t$:
> $x_f(t) = Mte^{-at}$.

The reason for this is that if the forcing function on the right-hand side of the differential equation has the form $f(t) = Ke^{-at}$, we would normally assume, by the method of undetermined coefficients, that the usual form for the forced solution is $x_f(t) = Me^{-at}$. But this would be identical to the *form* of the homogeneous solution, $x_h(t) = Ce^{-at}$, which satisfies the homogeneous equation

$$\frac{d}{dt}(Ce^{-at}) + a(Ce^{-at}) = 0$$

When we substitute the forced solution in order to determine the constant $M$ in that assumed form of the forced solution, we also obtain (remember that $K$ is known and is not zero)

$$\frac{d}{dt}(Me^{-at})+a(Me^{-at})= Ke^{-at}$$

or

$$-a(Me^{-at})+a(Me^{-at})=0$$
$$\neq Ke^{-at}$$

and we can find no solution for $M$! But assuming the forced solution to be $x_f(t) = Mte^{-at}$ would work, as shown in the following example.

---

### *Example*

Determine the solution to the following differential equation having the initial condition $x(0) = 2$:

$$\frac{dx(t)}{dt}+\underbrace{3}_{a}x(t)=\underbrace{10e^{-3t}}_{f(t)}\qquad x(0)=2$$

The homogeneous solution is

$$x_h(t) = Ce^{-3t}$$

Since $f(t) = 10e^{-3t}$, the usual form of the forced solution is

$$x_f(t) = Me^{-3t}$$

But since this is the same *form* as the homogeneous solution, we multiply the usual form of the forced solution by $t$ to give

$$x_f(t) = tMe^{-3t}$$

Substitute this form into the differential equation to give

$$(Me^{-3t} - 3tMe^{-3t})+3(tMe^{-3t})=10e^{-3t}$$

to give $M = 10$. Hence, the forced solution is

$$x_f(t) = 10te^{-3t}$$

The total solution is

$$x(t) = x_h(t) + x_f(t)$$
$$= Ce^{-3t} + 10te^{-3t}$$

Applying the initial condition gives

$$2 = Ce^{-3(0)}$$

so that $C = 2$, and hence the total solution valid for $t \geq 0$ is

$$x(t) = 2e^{-3t} + 10te^{-3t} \qquad \text{for} \quad t \geq 0$$

The reader should "sanity check" this result.

---

## 4.4  SOLUTION OF SECOND-ORDER EQUATIONS

Now we solve the second-order differential equation

$$\boxed{\frac{d^2x(t)}{dt^2} + a\frac{dx(t)}{dt} + bx(t) = f(t) \qquad x(0), \quad \dot{x}(0)} \tag{4.1b}$$

### 4.4.1  The Homogeneous Solution

The *homogeneous solution* satisfies

$$\boxed{\frac{d^2x_h(t)}{dt^2} + a\frac{dx_h(t)}{dt} + bx_h(t) = 0} \tag{4.14}$$

Once again, in order to satisfy this equation, the *form* of $x_h(t)$ must be such that its derivatives give back the same *form*. Again the only such form is the exponential

$$x_h(t) = Ce^{pt} \tag{4.15}$$

Substituting this into (4.14) gives

$$(p^2 + ap + b)Ce^{pt} = 0$$

Again, the only way that this can be satisfied is for the $p$ exponents to satisfy the *characteristic equation*:

$$\boxed{p^2 + ap + b = (p - p_1)(p - p_2) = 0}$$ (4.16)

whose roots are $p = p_1$ and $p = p_2$. This gives the general form of the homogeneous solution as

$$\boxed{x_h(t) = C_1 e^{p_1 t} + C_2 e^{p_2 t}}$$ (4.17)

The *two* undetermined constants $C_1$ and $C_2$ will be determined by applying the two initial conditions $x(0)$ and $\dot{x}(0)$ to the *total solution*.

The two roots of the characteristic equation, $p_1$ and $p_2$, can have three forms.

***Roots Real and Distinct:*** $\mathbf{p_1 \neq p_2}$   In this case the form of the homogeneous solution is given in (4.17):

$$\boxed{x_h(t) = C_1 e^{p_1 t} + C_2 e^{p_2 t}}$$ (4.17)

***Rools Real and Repeated:*** $\mathbf{p_1 = p_2 = p}$   If we use the form in (4.17) for this case, the two parts of the homogeneous solution will be identical and can therefore be combined, leaving only one undetermined constant. But we must have *two* undetermined constants in the homogeneous solution in order to evaluate the solution for the *two* initial conditions $x(0)$ and $\dot{x}(0)$. The general form of the homogeneous solution for this case is to multiply either $C_1 e^{pt}$ or $C_2 e^{pt}$ by $t$:

$$\boxed{x_h(t) = C_1 e^{pt} + C_2 t e^{pt}}$$ (4.18)

We can show that the second piece of the homogeneous solution, $C_1 t e^{pt}$, also satisfies the homogeneous differential equation for repeated roots. For repeated roots the characteristic equation becomes $p^2 + ap + b = (p - \bar{p})(p - \bar{p}) = p^2 - 2\bar{p}p + \bar{p}^2 = 0$. Here we have denoted the two roots as $p = \bar{p}$ to avoid confusion. We identify $a = -2\bar{p}$ and $b = \bar{p}^2$. Substituting $C_2 t e^{\bar{p}t}$ into (4.14) gives us

$$\frac{d^2}{dt}(C_2 t e^{\bar{p}t}) - 2\bar{p}\frac{d}{dt}(C_2 t e^{\bar{p}t}) + \bar{p}^2(C_2 t e^{\bar{p}t}) \overset{?}{=} 0$$

Performing the differentiations (using the "chain rule") gives us

$$(\bar{p}C_2 e^{\bar{p}t} + \bar{p}C_2 e^{\bar{p}t} + \bar{p}^2 C_2 t e^{\bar{p}t}) - 2\bar{p}(C_2 e^{\bar{p}t} + \bar{p}C_2 t e^{\bar{p}t}) + \bar{p}^2 C_2 t e^{\bar{p}t} \overset{?}{=} 0$$

Canceling terms, we see that this is satisfied.

What would you choose for the form of the homogeneous solution in the case of a third-order differential equation that has all three roots identical: $p_1 = p_2 = p_3 = p$?

***Roots Complex Conjugate:*** $\mathbf{p_1 = \alpha + j\beta}$, $\mathbf{p_2 = \alpha - j\beta}$    For complex-conjugate roots (which are distinct), there are several forms for the homogeneous solution, all of which have *two* undetermined constants:

$$\boxed{x_h(t) = C_1 e^{\alpha t} e^{j\beta t} + C_2 e^{\alpha t} e^{-j\beta t}} \tag{4.19a}$$

or

$$\boxed{x_h(t) = K_1 e^{\alpha t} \cos \beta t + K_2 e^{\alpha t} \sin \beta t} \tag{4.19b}$$

or

$$\boxed{x_h(t) = M_1 e^{\alpha t} \cos (\beta t + \theta_2)} \tag{4.19c}$$

or

$$\boxed{x_h(t) = N_1 e^{\alpha t} \sin (\beta t + \phi_2)} \tag{4.19d}$$

Applying the two initial conditions $x(0)$ and $\dot{x}(0)$ to the *total solution* gives the two undetermined constants $C_1$ and $C_2$, or $K_1$ and $K_2$, or $M_1$ and $\theta_2$, or $N_1$ and $\phi_2$, depending on which form you choose to use.

These equivalent forms are obtained using Euler's identity from Chapter 2:

$$e^{j\beta t} = 1\angle\beta t = \cos \beta t + j \sin \beta t$$

For example, (4.19b) can be obtained from (4.19a) by substituting Euler's identity to give

$$
\begin{aligned}
x_h(t) &= C_1 e^{\alpha t} e^{j\beta t} + C_2 e^{\alpha t} e^{-j\beta t} \\
&= C_1 e^{\alpha t}(\cos \beta t + j \sin \beta t) + C_2 e^{\alpha t}(\cos \beta t - j \sin \beta t) \\
&= \underbrace{(C_1 + C_2)}_{K_1} e^{\alpha t} \cos \beta t + \underbrace{j(C_1 - C_2)}_{K_2} e^{\alpha t} \sin \beta t
\end{aligned}
\tag{4.19a}
$$

Since $x_h(t)$ must be a real number (for physical systems), $K_1$ and $K_2$ must be real numbers. For this to be the case, the undetermined constants $C_1$ and $C_2$ in the original form in (4.19a) must be complex numbers and must be the conjugates of each other: $C_2 = C_1^*$, where * denotes the conjugate of a complex number. For example, if we write $C_1 = a + jb$ and $C_2 = C_1^* = a - jb$, then $K_1 = C_1 + C_2 = C_1 + C_1^* = 2a$ and $K_2 = j(C_1 - C_2) = j(C_1 - C_1^*) = -2b$, which are both real numbers. Hence $x_h(t)$ in (4.19a) and (4.19b) is a real number, as it should be in order to be the response of a physical system.

Similarly, (4.19c) can be obtained from (4.19a) by writing the undetermined coefficients in polar form as $C_1 = M\angle\theta = Me^{j\theta}$ and $C_2 = C_1^* = M\angle-\theta = Me^{-j\theta}$. Substituting gives

$$
\begin{aligned}
x_h(t) &= C_1 e^{\alpha t} e^{j\beta t} + C_2 e^{\alpha t} e^{-j\beta t} \\
&= C_1 e^{\alpha t} e^{j\beta t} + C_1^* e^{\alpha t} e^{-j\beta t} \\
&= Me^{j\theta} e^{\alpha t} e^{j\beta t} + Me^{-j\theta} e^{\alpha t} e^{-j\beta t} \\
&= Me^{\alpha t}(e^{j(\beta t+\theta)} + e^{-j(\beta t+\theta)}) \\
&= 2Me^{\alpha t}\cos(\beta t + \theta)
\end{aligned}
\tag{4.19a}
$$

which is the form in (4.19c). You should similarly verify the form in (4.19d). *Hint:* Write $C_1 = jN\angle\phi = jNe^{j\phi}$ and $C_2 = C_1^* = -jN\angle-\phi = -jNe^{-j\phi}$.

### 4.4.2  The Forced Solution for "Nice" $f(t)$

Determining the forced solution for a second-order differential equation is no different than for the first-order equation. Again use the "method of undetermined coefficients."

### 4.4.3  The Total Solution

The *total solution* is the sum of the homogeneous and forced solutions. We have three possibilities for these, for the three cases of the roots of the characteristic equation:

$$
\boxed{x(t) = C_1 e^{p_1 t} + C_2 e^{p_2 t} + x_f(t)}
\tag{4.20a}
$$

$$
\boxed{x(t) = C_1 e^{p t} + C_2 t e^{p t} + x_f(t)}
\tag{4.20b}
$$

$$
\boxed{x(t) = C_1 e^{\alpha t}\cos\beta t + C_2 e^{\alpha t}\sin\beta t + x_f(t)}
\tag{4.20c}
$$

Applying the two initial conditions, $x(0)$ and its derivative at $t = 0$, $dx(t)/dt|_{t=0} \equiv \dot{x}(0)$, gives two simultaneous algebraic equations to solve for $C_1$ and $C_2$ for these three cases:

$$
\boxed{
\begin{aligned}
x(0) &= C_1 + C_2 + x_f(0) \\
\dot{x}(0) &= p_1 C_1 + p_2 C_2 + \dot{x}_f(0)
\end{aligned}
}
\tag{4.21a}
$$

$$
\boxed{
\begin{aligned}
x(0) &= C_1 + x_f(0) \\
\dot{x}(0) &= p C_1 + C_2 + \dot{x}_f(0)
\end{aligned}
}
\tag{4.21b}
$$

$$\boxed{\begin{aligned} x(0) &= C_1 + x_f(0) \\ \dot{x}(0) &= \alpha C_1 + \beta C_2 + \dot{x}_f(0) \end{aligned}} \qquad (4.21c)$$

which you should verify. See Figure 1.4 for plots of these three forms of the solution.

---

### Example

Determine the general solution for the differential equation

$$\frac{d^2 x(t)}{dt} + 3\frac{dx(t)}{dt} + 2x(t) = 10$$

The characteristic equation is

$$p^2 + 3p + 2 = (p+1)(p+2) = 0$$

(which you should "sanity check"). The two roots are $p_1 = -1$ and $p_2 = -2$. The form of the homogeneous solution is

$$x_h(t) = C_1 e^{-t} + C_2 e^{-2t}$$

Since $f(t) = 10$, we assume the form of the forced solution to be $x_f(t) = M$ and substitute to give

$$x_f(t) = 5$$

Hence, the form of the total solution is

$$x(t) = C_1 e^{-t} + C_2 e^{-2t} + 5$$

Substituting the two initial conditions gives the two simultaneous algebraic equations in (4.21a) to solve for the two undetermined coefficients $C_1$ and $C_2$ (as you should verify):

$$x(0) = C_1 + C_2 + 5$$
$$\dot{x}(0) = -C_1 - 2C_2$$

thereby completing the solution.

## *Example*

Determine the general solution for the differential equation

$$\frac{d^2x(t)}{dt^2} + 6\frac{dx(t)}{dt} + 9x(t) = 2t$$

The characteristic equation is

$$p^2 + 6p + 9 = (p+3)(p+3) = 0$$

(which you should "sanity check"). The two roots are repeated: $p_1 = -3$ and $p_2 = -3$. The form of the homogeneous solution is

$$x_h(t) = C_1e^{-3t} + C_2te^{-3t}$$

Since $f(t) = 2t$, we assume the form of the forced solution to be $x_f(t) = M_1 + M_2t$ and substitute to give (as you should verify)

$$x_f(t) = -\frac{4}{27} + \frac{2}{9}t$$

Hence, the form of the total solution is

$$x(t) = C_1e^{-3t} + C_2te^{-3t} - \frac{4}{27} + \frac{2}{9}t$$

Substituting the two initial conditions gives the two simultaneous algebraic equations in (4.21b) to solve for the two undetermined coefficients $C_1$ and $C_2$ (as you should verify):

$$x(0) = C_1 - \frac{4}{27}$$

$$\dot{x}(0) = -3C_1 + C_2 + \frac{2}{9}$$

thereby completing the solution.

### Example

Determine the general solution for the differential equation

$$\frac{d^2x(t)}{dt} + 4\frac{dx(t)}{dt} + 13x(t) = 10$$

The characteristic equation is

$$p^2 + 4p + 13 = (p + 2 + j3)(p + 2 - j3) = 0$$

(which you should "sanity check"). The two roots are $p_1 = -2 - j3$ and $p_2 = -2 + j3$. One form of the homogeneous solution is

$$x_h(t) = C_1 e^{-2t}\cos 3t + C_2 e^{-2t}\sin 3t$$

There are other possible *forms* for the homogeneous solution, as shown in (4.19). Since $f(t) = 10$, we assume the form of the forced solution to be $x_f(t) = M$ and substitute to give

$$x_f(t) = \frac{10}{13}$$

Hence the form of the total solution is

$$x(t) = C_1 e^{-2t}\cos 3t + C_2 e^{-2t}\sin 3t + \frac{10}{13}$$

Substituting the two initial conditions gives the two simultaneous algebraic equations in (4.21c) to solve for the two undetermined coefficients $C_1$ and $C_2$ (as you should verify):

$$x(0) = C_1 + \frac{10}{13}$$

$$\dot{x}(0) = -2C_1 + 3C_2$$

thereby completing the solution.

### 4.4.4  A Special Case

If the forcing function $f(t)$ has the same *form* as any part of the homogeneous solution, we cannot use the usual assumed form of the forced solution $x_f(t)$

because that would be identical to the *form of the homogeneous solution that satisfies the homogeneous differential equation whose right-hand side is zero.* Hence, we would not be able to determine the unknowns in this assumed form of the forced solution by substituting it into the differential equation representing the forced solution whose right-hand side is *not zero. If the forcing function f(t) has the same form as any part of the homogeneous solution, we must multiply the usual assumed form for the forced solution by either t (for distinct roots) or t² (for repeated roots).*

### Distinct Roots (Which Includes Complex Roots): $p_1 \neq p_2$

$$\boxed{\begin{aligned} f(t) &= Ke^{p_1 t} \Rightarrow x_f(t) = Mte^{p_1 t} \\ f(t) &= Ke^{p_2 t} \Rightarrow x_f(t) = Mte^{p_2 t} \end{aligned}}$$

(4.22a)

---

### *Example*

For the differential equation

$$\frac{d^2 x(t)}{dt^2} + 3\frac{dx(t)}{dt} + 2x(t) = 5e^{-2t}$$

the characteristic equation is

$$p^2 + 3p + 2 = (p+1)(p+2) = 0$$

giving the two roots as $p_1 = -1$ and $p_2 = -2$. The homogeneous solution is

$$x_h(t) = C_1 e^{-t} + C_2 e^{-2t}$$

For $f(t) = 5e^{-2t}$ the usual form of the forced solution would be $x_f(t) = Me^{-2t}$. But that would be the same form as the homogeneous solution $C_2 e^{-2t}$, which satisfies the *homogeneous equation* whose right-hand side is *not zero.* So we choose

$$x_f(t) = Mte^{-2t}$$

Substituting into the differential equation gives

$$(-2Me^{-2t} - 2Me^{-2t} + 4Mte^{-2t}) + 3(Me^{-2t} - 2Mte^{-2t}) + 2(Mte^{-2t}) = 5e^{-2t}$$

Canceling terms gives

$$-Me^{-2t} = 5e^{-2t}$$

giving $M = -5$, and the forced solution is

$$x_f(t) = -5te^{-2t}$$

---

### Repeated Roots: $p_1 = p_2 = p$

$$\boxed{f(t) = Ke^{pt} \Rightarrow x_f(t) = Mt^2e^{pt}}$$     (4.22b)

---

### Example

For the differential equation

$$\frac{d^2x(t)}{dt^2} + 4\frac{dx(t)}{dt} + 4x(t) = 5e^{-2t}$$

the characteristic equation is

$$p^2 + 4p + 4 = (p+2)(p+2) = 0$$

giving the two roots as $p_1 = p_2 = -2$. The homogeneous solution is

$$x_h(t) = C_1e^{-2t} + C_2te^{-2t}$$

For $f(t) = 5e^{-2t}$ the usual form of the forced solution would be $x_f(t) = Me^{-2t}$. But that would be the same form as the homogeneous solution $C_1e^{-2t}$, which satisfies the *homogeneous equation* whose right-hand side is *not zero*. Assuming the form $x_f(t) = Mte^{-2t}$ would not work either since it would be the same form as the other part of the homogeneous solution $C_2te^{-2t}$, which also satisfies the *homogeneous equation* whose right-hand side is *not zero*. So we choose

$$x_f(t) = Mt^2e^{-2t}$$

Substituting into the differential equation gives

$$(2Me^{-2t} - 4Mte^{-2t} - 4Mte^{-2t} + 4Mt^2e^{-2t})$$
$$+ 4(2Mte^{-2t} - 2Mt^2e^{-2t}) + 4(Mt^2e^{-2t}) = 5e^{-2t}$$

Canceling terms gives

$$2Me^{-2t} = 5e^{-2t}$$

giving $M = 5/2$, and the forced solution is

$$x_f(t) = \frac{5}{2}t^2 e^{-2t}$$

---

## 4.5  STABILITY OF THE SOLUTION

In physical systems, as time increases, $t \to \infty$, the homogeneous solution should go to zero:

$$\lim_{t \to \infty} x_h(t) = 0$$

so that as $t \to \infty$, the solution approaches the forced solution:

$$\lim_{t \to \infty} x(t) = \lim_{t \to \infty} [x_h(t) + x_f(t)] = x_f(t)$$

Hence, the forced solution is often called the *steady-state solution*.

Since the homogeneous solution is a sum of exponentials, this will be the case only if the roots of the characteristic equation, $p_1$ and $p_2$, have negative real parts. For a second-order equation the roots of the characteristic equation are

$$p_1, p_2 = -\frac{a}{2} \pm \frac{1}{2}\sqrt{a^2 - 4b}$$
$$= \alpha \pm j\beta$$

The form of the homogeneous solution is

$$x_h(t) = C_1 e^{p_1 t} + C_2 e^{p_2 t}$$
$$= C_1 e^{\alpha t} e^{j\beta t} + C_2 e^{\alpha t} e^{-j\beta t}$$

So for the solution to be *stable*, the real parts of the roots, $\alpha$, must be negative; otherwise, the total solution will "blow up" as $t \to \infty$. This would not be a sensible solution for a physical system since an infinite amount of energy would be required to be delivered by the system. The above also applies to the cases of real, distinct roots, and real, repeated roots: Both roots must be negative for physical systems that are governed by the differential equation.

## 4.6  SOLUTION OF SIMULTANEOUS SETS OF ORDINARY DIFFERENTIAL EQUATIONS WITH THE DIFFERENTIAL OPERATOR

The differential operator $D$ "operates on" a function, $f(t)$, to yield the derivative of that function:

$$\boxed{D^n f(t) \Rightarrow \frac{d^n f(t)}{dt^n}}$$    (4.23)

The differential operator obeys the rules of algebra and can be treated like an algebraic quantity:

$$D^n[f(t) + g(t)] = \frac{d^n f(t)}{dt^n} + \frac{d^n g(t)}{dt^n}$$    (4.24a)

$$D^m D^n f(t) = D^{m+n} f(t) = \frac{d^{m+n} f(t)}{dt^{m+n}}$$    (4.24b)

$$\frac{1}{D^n} D^m f(t) = D^{m-n} f(t) = \frac{d^{m-n} f(t)}{dt^{m-n}}$$    (4.24c)

$$\frac{1}{D^n} f(t) = \underbrace{\int \cdots \int f(\tau) d\tau}_{n}$$    (4.24d)

The $D$ operator *must appear on the left of the function* or it wouldn't make sense. For example, writing $f(t)D$ wouldn't make sense because there is no function on the right of $D$ for it to "operate on."

The differential operator can be used to write differential equations in a tidy, symbolic form. One of the primary advantages of doing this with the differential operator is that *simultaneous sets* of differential equations can be manipulated and solved easily, as the following examples show.

---

### Example

Write the following set of two simultaneous differential equations in symbolic form using the differential operator:

$$3\frac{d^2 x(t)}{dt^2} + 2\frac{dx(t)}{dt} + 5\frac{dy(t)}{dt} + y(t) = 10\sin 4t$$

$$\frac{dx(t)}{dt} - 3x(t) + 2\frac{d^2 y(t)}{dt^2} = 5t^2$$

Replacing derivatives with the differential operator gives the symbolic differential equations

$$(3D^2 + 2D)x(t) + (5D+1)y(t) = 10\sin 4t$$

$$(D-3)x(t) + (2D^2)y(t) = 5t^2$$

You should check this by writing it out.

---

The problem with simultaneous sets of differential equations is that each differential equation in the set involves more than one variable. For example, each equation in the simultaneous differential equations of the preceding example contains *both* $x(t)$ and $y(t)$. The two equations in the set are therefore said to be *coupled*. To solve the simultaneous set, we must be able to reduce them to two *uncoupled* differential equations each involving *only* one variable, such as $x(t)$ or $y(t)$. The differential operator reduces the coupled set of *differential equations* to a coupled set of *algebraic equations* which we know how to solve (Chapter 3). These can then be manipulated to obtain uncoupled differential equations. The following example illustrates this valuable property of the differential operator.

---

### *Example*

Write the following *coupled* differential equations using the differential operator as in the preceding example:

$$\frac{dx(t)}{dt} + x(t) + y(t) = 2\sin 3t$$

$$-2x(t) + \frac{dy(t)}{dt} + 4y(t) = 5t^2$$

$$\Downarrow$$

$$(D+1)x(t) + y(t) = 2\sin 3t$$

$$-2x(t) + (D+4)y(t) = 5t^2$$

Reduce them to two differential equations, each of which contains only one unknown; $x(t)$ or $y(t)$, but not both. To do so, we may write these in matrix form (Chapter 3) as

$$\begin{bmatrix} D+1 & 1 \\ -2 & D+4 \end{bmatrix}\begin{bmatrix} x(t) \\ y(t) \end{bmatrix} = \begin{bmatrix} 2\sin 3t \\ 5t^2 \end{bmatrix}$$

Solving by Cramer's rule gives

$$x(t) = \frac{\begin{vmatrix} 2\sin 3t & 1 \\ 5t^2 & D+4 \end{vmatrix}}{\begin{vmatrix} D+1 & 1 \\ -2 & D+4 \end{vmatrix}} = \frac{(D+4)2\sin 3t - 5t^2}{D^2 + 5D + 6} = \frac{6\cos 3t + 8\sin 3t - 5t^2}{D^2 + 5D + 6}$$

$$y(t) = \frac{\begin{vmatrix} D+1 & 2\sin 3t \\ -2 & 5t^2 \end{vmatrix}}{\begin{vmatrix} D+1 & 1 \\ -2 & D+4 \end{vmatrix}} = \frac{(D+1)5t^2 + 4\sin 3t}{D^2 + 5D + 6} = \frac{10t + 5t^2 + 4\sin 3t}{D^2 + 5D + 6}$$

Multiplying both sides by the denominator gives the individual differential equations as

$$(D^2 + 5D + 6)x(t) = 6\cos 3t + 8\sin 3t - 5t^2$$

$$(D^2 + 5D + 6)y(t) = 10t + 5t^2 + 4\sin 3t$$

or

$$\frac{d^2x(t)}{dt^2} + 5\frac{dx(t)}{dt} + 6x(t) = 6\cos 3t + 8\sin 3t - 5t^2$$

$$\frac{d^2y(t)}{dt^2} + 5\frac{dy(t)}{dt} + 6y(t) = 10t + 5t^2 + 4\sin 3t$$

These individual differential equations can be solved by the previous methods, giving the total solutions as

$$x(t) = A_1 e^{-2t} + A_2 e^{-3t} - \frac{23}{39}\cos 3t + \frac{11}{39}\sin 3t - \frac{95}{108} + \frac{25}{18}t - \frac{5}{6}t^2$$

$$y(t) = B_1 e^{-2t} + B_2 e^{-3t} - \frac{10}{39}\cos 3t - \frac{2}{39}\sin 3t - \frac{55}{108} + \frac{5}{18}t + \frac{5}{6}t^2$$

which you should verify.

The four undetermined constants in the homogeneous solutions for the coupled differential equations in the preceding example are related.

*Example*

For example, write the first differential equation,

$$\frac{dx(t)}{dt} + x(t) + y(t) = 2\sin 3t$$

as

$$y(t) = -\frac{dx(t)}{dt} - x(t) + 2\sin 3t$$

Write the solution for $x(t)$ as

$$x(t) = \underbrace{A_1 e^{-2t} + A_2 e^{-3t}}_{x_h(t)} + x_f(t)$$

Substitute to give

$$y(t) = -\frac{dx(t)}{dt} - x(t) + 2\sin 3t$$

$$= 2A_1 e^{-2t} + 3A_2 e^{-3t} - A_1 e^{-2t} - A_2 e^{-3t}$$

$$\qquad - \frac{dx_f(t)}{dt} - x_f(t) + 2\sin 3t$$

$$= A_1 e^{-2t} + 2A_2 e^{-3t} - \frac{dx_f(t)}{dt} - x_f(t) + 2\sin 3t$$

Comparing this to the solution for $y(t)$,

$$y(t) = \underbrace{B_1 e^{-2t} + B_2 e^{-3t}}_{y_h(t)} + y_f(t)$$

we see that the undetermined coefficients in the homogeneous solution for $y(t)$ are related to those in the homogeneous solution for $x(t)$ as

$$B_1 = A_1$$

$$B_2 = 2A_2$$

Hence, *the number of initial conditions that can be specified is equal to the order of the determinant in the denominator of Cramer's rule:* $D^2 + 5D + 6$. We could specify

$$x(0) \quad \text{and} \quad \left.\frac{dx(t)}{dt}\right|_{t=0}$$

or

$$y(0) \quad \text{and} \quad \left.\frac{dy(t)}{dt}\right|_{t=0}$$

or

$$x(0) \quad \text{and} \quad \left.\frac{dy(t)}{dt}\right|_{t=0}$$

or

$$y(0) \quad \text{and} \quad \left.\frac{dx(t)}{dt}\right|_{t=0}$$

### 4.6.1    Using the Differential Operator to Verify Solutions

The differential operator can also be used to verify, *easily* and *immediately*, the various forms of the homogeneous and forced solutions. First write the second-order differential equation

$$\frac{d^2x(t)}{dt^2} + a\frac{dx(t)}{dt} + bx(t) = f(t)$$

using the differential operator as

$$(D^2 + aD + b)x(t) = f(t)$$

Write this in factored form as

$$\boxed{(D - p_1)(D - p_2)x(t) = f(t)} \tag{4.25}$$

where $p_1$ and $p_2$ are the roots of the characteristic equation. The important functions that appear in the various solutions are the exponentials $e^{pt}$, $te^{pt}$, $t^2e^{pt}$, and so on. First we establish some important results:

$$\boxed{(D - p)e^{pt} = pe^{pt} - pe^{pt} = 0} \tag{4.26a}$$

$$\boxed{(D - p)te^{pt} = e^{pt} + pte^{pt} - pte^{pt} = e^{pt}} \tag{4.26b}$$

$$\boxed{(D - p)t^2e^{pt} = 2te^{pt} + pt^2e^{pt} - pt^2e^{pt} = 2te^{pt}} \tag{4.26c}$$

$$\boxed{(D - p)t^3e^{pt} = 3t^2e^{pt} + pt^3e^{pt} - pt^3e^{pt} = 3t^2e^{pt}} \tag{4.26d}$$

The differentiations in (4.26b), (4.26c), and (4.26d) were obtained using the chain rule (Chapter 2).

Using these fundamental results, we are now able to verify easily that the previously determined forms of the homogeneous and forced solutions indeed satisfy the appropriate differential equation.

***Homogeneous Solution* f(t) = 0, *Distinct Roots:* $p_1 \neq p_2$**    The homogeneous solution for this case is the sum of the two exponential functions $C_1e^{p_1t}$ and $C_2e^{p_2t}$. Substituting $C_1e^{p_1t}$ into (4.25) gives

$$(D - p_1)(D - p_2)C_1e^{p_1t} \overset{?}{=} 0$$

Interchanging the operators and using (4.26a) shows immediately that

$$(D - p_2)\underbrace{(D - p_1)C_1e^{p_1t}}_{0} \overset{?}{=} 0$$

Similarly, substituting $C_2 e^{p_2 t}$ into (4.25) and using (4.26a) shows immediately that

$$(D - p_1)\underbrace{(D - p_2)C_2 e^{p_2 t}}_{0} \overset{?}{=} 0$$

So the differential operator is a very simple means of demonstrating immediately that these assumed forms in fact satisfy the homogeneous equation.

***Homogeneous Solution f(t) = 0, Repeated Roots: $p_1 = p_2 = p$***    The homogeneous solution for this case is the sum of two exponential functions, $C_1 e^{pt}$ and $C_2 t e^{pt}$. Substituting $C_1 e^{pt}$ into (4.25) gives

$$(D - p)(D - p)C_1 e^{pt} \overset{?}{=} 0$$

Using (4.26a) immediately shows that

$$(D - p)\underbrace{(D - p)C_1 e^{pt}}_{0} \overset{?}{=} 0$$

Similarly, substituting $C_2 t e^{pt}$ into (4.25) and using (4.26b) gives

$$(D - p)\underbrace{(D - p)C_2 t e^{pt}}_{C_2 e^{pt}} \overset{?}{=} 0$$

Operating on this with the second operator gives, using (4.26a),

$$\underbrace{(D - p)C_2 e^{pt}}_{0} = 0$$

Again the differential operator has provided an easy and immediate means of demonstrating that these assumed forms in fact satisfy the homogeneous equation.

***The Forced Solution f(t) ≠ 0 for Special Cases***    Determining the forced solution is particularly simple using the *method of undetermined coefficients:* Assume a *form* of the forced solution that is the same *form* as $f(t)$. But there were two special cases where this method required modification. The first case was where the roots of the characteristic equation are distinct and $f(t) = K e^{p_i t}$, with $p_i$ being a root of the characteristic equation. If we assume the usual form of the forced solution as $x_f(t) = M e^{p_i t}$, this could not be made to satisfy the nonhomogeneous equation because that form was the same form as in the homogeneous solution. We can easily demonstrate this with the differential operator

$$(D - p_2)\underbrace{(D - p_1)\,Me^{p_1 t}}_{0} \neq Ke^{p_1 t}$$

$$(D - p_1)\underbrace{(D - p_2)\,Me^{p_2 t}}_{0} \neq Ke^{p_2 t}$$

where we used the result in (4.26a). For this case we multiplied the usual assumed form of the forced solution by $t$ to give $x_f(t) = Mte^{p_1 t}$. Using the differential operator, we can easily show that this form can be made to satisfy the nonhomogeneous equation

$$(D - p_2)\underbrace{(D - p_1)\,Mte^{p_1 t}}_{Me^{p_1 t}} = (p_1 - p_2)\,Me^{p_1 t} = Ke^{p_1 t}$$

$$(D - p_1)\underbrace{(D - p_2)\,Mte^{p_2 t}}_{Me^{p_2 t}} = (p_2 - p_1)\,Me^{p_2 t} = Ke^{p_2 t}$$

where we used the result in (4.26b), thereby giving a solution for $M$ in either case.

If the roots of the characteristic equation are repeated, $p_1 = p_2 = p$ and $f(t) = Ke^{pt}$, assuming a form of the forced solution $x_f(t) = Mte^{pt}$ would not satisfy the nonhomogeneous equation since $p_1 = p_2$ and the left-hand sides of the previous result would be zero. This is again rather obvious because $x_f(t) = Mte^{pt}$ has the same form as a part of the homogeneous solution for this case of repeated roots and satisfies the homogeneous equation whose right-hand side is zero. So we again multiplied this by $t$ and assumed the form of the forced solution to be $x_f(t) = Mt^2 e^{pt}$. Using the differential operator, we can easily show that this form can be made to satisfy the nonhomogeneous equation

$$(D - p)\underbrace{(D - p)\,Mt^2 e^{pt}}_{2Mte^{pt}} = 2Me^{pt} = Ke^{pt}$$

where we used the basic result in (4.26c) and then the basic result in (4.26b).

To illustrate the power of this method, suppose that the roots of the characteristic equation are repeated, $p_1 = p_2 = p$, and the forcing function $f(t)$ has the form $f(t) = Kte^{pt}$ (a case not covered in Section 4.4.2 or Section 4.4.4). An assumed form of $x_f(t) = Mt^2 e^{pt}$ could not be made to satisfy the nonhomogeneous equation

$$(D - p)\underbrace{(D - p)\,Mt^2 e^{pt}}_{2Mte^{pt}} = 2Me^{pt} \neq Kte^{pt}$$

This is not so obvious because the homogeneous solution does not contain a term of this form. But the form $x_f(t) = Mt^3 e^{pt}$ would work:

$$(D-p)\underbrace{(D-p)\,Mt^3e^{pt}}_{3Mt^2e^{pt}} = 6Mte^{pt} = Kte^{pt}$$

## 4.7  NUMERICAL (COMPUTER) SOLUTIONS

Ordinary differential equations (even nonlinear ones) can be solved (approximately) using a digital computer. The solution process, although approximate, is *very simple*! For example, a first-order, linear, constant-coefficient, ordinary differential equation is

$$\frac{dy(t)}{dt} + ay(t) = f(t)$$

where $a$ and $f(t)$ are known (given) along with the initial condition $y(0)$. To solve this approximately, we divide the $t$ axis into equal-length segments $\Delta t$ and obtain the solution only at the points $0, \Delta t, 2\Delta t, 3\Delta t, \ldots$, as illustrated in Figure 4.2. Approximate the derivative as

$$\left.\frac{dy(t)}{dt}\right|_{n\Delta t} \cong \frac{y((n+1)\Delta t) - y(n\Delta t)}{\Delta t} = \frac{y_{n+1} - y_n}{\Delta t}$$

where we denote

$$y(n\Delta t) \equiv y_n$$

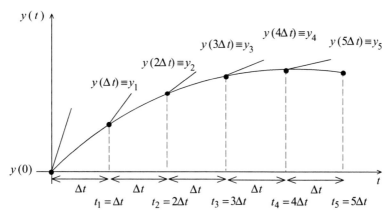

**Fig. 4.2.** Approximating a function over discrete time intervals of $\Delta t$.

This is essentially an *approximation* to the *slope* of the function at $t_n \equiv n\Delta t$. Rewriting this as

$$y_{n+1} = y_n + \Delta t \left. \frac{dy(t)}{dt} \right|_{n\Delta t}$$

shows that we are *predicting* the next solution point, $y_{n+1}$, from the previous point, $y_n$, and the *slope* of the curve at the previous point, $y_n$. Note that for this to be a good approximation, the points $0$, $\Delta t$, $2\Delta t$, $3\Delta t$, ... must be "close together." The more rapidly the solution $y(t)$ is changing, the closer together the points must be to give a good approximation.

Substitute the foregoing approximation for the derivative into the differential equation

$$\frac{y_{n+1} - y_n}{\Delta t} + ay_n = f(t_n) \equiv f_n$$

Multiplying both sides by $\Delta t$ gives

$$y_{n+1} - y_n + a\,\Delta t y_n = \Delta t f_n$$

Solving gives a recursion relation:

$$\boxed{y_{n+1} = (1 - a\,\Delta t)\,y_n + \Delta t f_n} \tag{4.27}$$

So we can solve this in a recursive or "marching in time" fashion:

$$y_1 = (1 - a\,\Delta t)\,y_0 + \Delta t f_0$$
$$y_2 = (1 - a\,\Delta t)\,y_1 + \Delta t f_1$$
$$y_3 = (1 - a\,\Delta t)\,y_2 + \Delta t f_2$$
$$y_4 = (1 - a\,\Delta t)\,y_3 + \Delta t f_3$$
$$\vdots$$

We know the initial condition, $y(0)$, and $f(t)$, so we can get $f(0)$, $f(\Delta t)$, $f(2\Delta t)$, $f(3\Delta t)$, ... . These are a form of difference equations that we investigate in Chapter 5. Solve these "recursively" by solving the first equation for $y_1 \equiv y(\Delta t)$. Substitute this into the second equation and solve for $y_2 \equiv y(2\Delta t)$, and so on. This is called the *forward Euler method*. The stability of the method is not very good; that is, if the step size, $\Delta t$, is not "small enough," the result may start to "diverge" from the true solution as we continue.

The *backward Euler method* is a better method and is absolutely stable no matter the step size. It is obtained from

$$\frac{y_{n+1} - y_n}{\Delta t} + ay_{n+1} = f(t_{n+1}) \equiv f_{n+1}$$

(What is the difference between the forward and backward Euler discretizations?) Solving this gives the recursion relation

$$\boxed{y_{n+1} = \frac{1}{1+a\,\Delta t}\,y_n + \frac{\Delta t}{1+a\,\Delta t}\,f_{n+1}}$$                                (4.28)

as you should verify. The recursion relations are

$$y_1 = \frac{1}{1+a\,\Delta t}\,y_0 + \frac{\Delta t}{1+a\,\Delta t}\,f_1$$

$$y_2 = \frac{1}{1+a\,\Delta t}\,y_1 + \frac{\Delta t}{1+a\,\Delta t}\,f_2$$

$$y_3 = \frac{1}{1+a\,\Delta t}\,y_2 + \frac{\Delta t}{1+a\,\Delta t}\,f_3$$

$$y_4 = \frac{1}{1+a\,\Delta t}\,y_3 + \frac{\Delta t}{1+a\,\Delta t}\,f_4$$

This backward Euler method is absolutely stable and will not diverge, no matter the step size.

The recursion relations for either method can be implemented easily using Excel, MATLAB, or your own calculator.

---

***Example***

$$\frac{dy(t)}{dt} + 2y(t) = 4 \qquad y(0) = 3$$

The exact solution is ("sanity check" it)

$$y(t) = e^{-2t} + 2 \qquad \text{for } t \geq 0$$

Comparing this to the general form we are considering, we identify

$$a = 2 \qquad f(t) = 4$$

If we choose a step size of $\Delta t = 1$, the recursion relations for the forward Euler method are

$$y_1 = -y_0 + 4$$
$$y_2 = -y_1 + 4$$

$$y_3 = -y_2 + 4$$

$$y_4 = -y_3 + 4$$

$$\vdots$$

and for the backward Euler method are

$$y_1 = \frac{1}{3} y_0 + \frac{4}{3}$$

$$y_2 = \frac{1}{3} y_1 + \frac{4}{3}$$

$$y_3 = \frac{1}{3} y_2 + \frac{4}{3}$$

$$y_4 = \frac{1}{3} y_3 + \frac{4}{3}$$

$$\vdots$$

The results for $\Delta t = 1$ are:

| $y_n$ | Exact | Forward | Backward |
|---|---|---|---|
| $\Delta t = 1$ | 2.135 | 1 | 2.333 |
| $2\Delta t = 2$ | 2.018 | 3 | 2.111 |
| $3\Delta t = 3$ | 2.002 | 1 | 2.037 |
| $4\Delta t = 4$ | 2.000 | 3 | 2.012 |

The backward Euler results are converging to the true steady-state value of 2, but the forward Euler results are "oscillating" about it.

You can check to see that if we had chosen $\Delta t > 1$, the forward Euler solution would increase without bound, that is, would "blow up," meaning that the solution process is unstable for $\Delta t \geq 1$! The backward Euler solution would, however, converge to 2 and is therefore stable.

If we reduce the step size to $\Delta t = 0.1$, the recursion relations for the forward Euler method are

$$y_1 = 0.8y_0 + 0.4$$

$$y_2 = 0.8y_1 + 0.4$$

$$y_3 = 0.8y_2 + 0.4$$

$$y_4 = 0.8y_3 + 0.4$$

$$\vdots$$

and for the backward Euler method are

$$y_1 = \frac{5}{6} y_0 + \frac{1}{3}$$

$$y_2 = \frac{5}{6} y_1 + \frac{1}{3}$$

$$y_3 = \frac{5}{6} y_2 + \frac{1}{3}$$

$$y_4 = \frac{5}{6} y_3 + \frac{1}{3}$$

$$\vdots$$

The results for $\Delta t = 0.1$ are:

| $y_n$ | Exact | Forward | Backward |
|---|---|---|---|
| $\Delta t = 0.1$ | 2.819 | 2.800 | 2.833 |
| $2\Delta t = 0.2$ | 2.670 | 2.640 | 2.694 |
| $3\Delta t = 0.3$ | 2.549 | 2.512 | 2.579 |
| $4\Delta t = 0.4$ | 2.449 | 2.410 | 2.482 |
| $5\Delta t = 0.5$ | 2.368 | 2.328 | 2.402 |
| $6\Delta t = 0.6$ | 2.301 | 2.262 | 2.335 |
| $7\Delta t = 0.7$ | 2.247 | 2.210 | 2.279 |
| $8\Delta t = 0.8$ | 2.202 | 2.168 | 2.233 |
| $9\Delta t = 0.9$ | 2.165 | 2.134 | 2.194 |
| $10\Delta t = 1.0$ | 2.135 | 2.107 | 2.162 |

Both methods are converging to the true steady-state value of 2.

There are other stable and more accurate methods for these and higher-order ordinary differential equations, such as the trapezoidal method and the Runge–Kutta method, but the basic idea is the same: Discretize the $t$ axis into $\Delta t$ increments and solve the resulting recursion relation in a "marching in time" manner.

# 5 Solution of Linear, Constant-Coefficient, Difference Equations

The governing equations of most physical systems generally describe the behavior of the system as a function of *time, t*. In Chapter 4 we investigated the solution of the ordinary differential equations that describe (govern) the behavior of a system as a *continuous* function of time. These are said to be *continuous-time systems*. There are numerous other physical systems, such as digital systems wherein the system behavior happens at *discrete* points in time. These are said to be *discrete-time systems*. These types of systems are described by *difference equations*, whose solution we study in this chapter.

## 5.1 WHERE DIFFERENCE EQUATIONS ARISE

There are numerous examples of *discrete-time systems* that are described by *difference equations*. As a familiar example, in Section 4.7 we examined the solution of an ordinary differential equation in an approximate fashion by discretizing the time axis into $\Delta t$ increments. We then developed a *recursion relation* which when solved iteratively gives the solution at discrete time intervals of $t = 0, t = \Delta t, t = 2\Delta t, \ldots$ . For example, we solved the first-order, ordinary differential equation

$$\frac{dy(t)}{dt} + ay(t) = f(t)$$

by discretizing the time axis into equally spaced points separated by $\Delta t$, as $t = 0, \Delta t, 2\Delta t, 3\Delta t, \ldots$ . We approximated the derivative as

$$\frac{dy(t)}{dt}\bigg|_{n\Delta t} \cong \frac{y((n+1)\Delta t) - y(n\Delta t)}{\Delta t} = \frac{y_{n+1} - y_n}{\Delta t}$$

where we denote

$$y(n\Delta t) \equiv y_n$$

*Essential Math Skills for Engineers*, By Clayton R. Paul
Copyright © 2009 John Wiley & Sons, Inc.

Subtituting this approximation for the derivative into the differential equation gives

$$\frac{y_{n+1} - y_n}{\Delta t} + a y_n = f_n$$

where we denote

$$f(n\Delta t) \equiv f_n$$

Solving gives a *recursion relation*:

$$y_{n+1} = (1 - a\Delta t)y_n + \Delta t\, f_n$$

We can solve this in a recursive or "marching in time" fashion:

$$y_1 = (1 - a\Delta t)y_0 + \Delta t\, f_0$$
$$y_2 = (1 - a\Delta t)y_1 + \Delta t\, f_1$$
$$y_3 = (1 - a\Delta t)y_2 + \Delta t\, f_2$$
$$y_4 = (1 - a\Delta t)y_3 + \Delta t\, f_3$$
$$\vdots$$

We know the initial condition, $y(0)$, and $f(t)$, so we can get $f(0)$, $f(\Delta t)$, $f(2\Delta t)$, $f(3\Delta t)$, .... . Solve these "recursively" by solving the first equation for $y_1 \equiv y(\Delta t)$. Substitute this into the second equation and solve for $y_2 \equiv y(2\Delta t)$, and so on. This is called the forward Euler method. We saw that in order to obtain an accurate and stable solution, the time interval discretization, $\Delta t$, must be chosen "small." So to obtain the solution at a distant time, say for large $n$, we must solve these for a large number of iterations. We found that to obtain a solution that is *stable* (doesn't "blow up"), the $\Delta t$ must be chosen "small enough" such that $|1 - a\Delta t| \leq 1$. Therefore, we must choose $\Delta t$ small enough to ensure stability of the solution. In addition, to obtain enough solution points to give sufficient detail of the solution, we must choose $\Delta t$ so small that we will require a very large number of iterations to get to the final time at which we desire a solution. So for this method we generally will have to iterate for a very large number of steps.

We also obtained another recursion relation using the backward Euler method:

$$\frac{y_{n+1} - y_n}{\Delta t} + a y_{n+1} = f(t_{n+1}) \equiv f_{n+1}$$

This is solved to give the recursion relation for the backward Euler method:

$$y_{n+1} = \frac{1}{1 + a\Delta t} y_n + \frac{\Delta t}{1 + a\Delta t} f_{n+1}$$

These can also be solved recursively as

$$y_1 = \frac{1}{1+a\,\Delta t}\,y_0 + \frac{\Delta t}{1+a\,\Delta t}\,f_1$$

$$y_2 = \frac{1}{1+a\,\Delta t}\,y_1 + \frac{\Delta t}{1+a\,\Delta t}\,f_2$$

$$y_3 = \frac{1}{1+a\,\Delta t}\,y_2 + \frac{\Delta t}{1+a\,\Delta t}\,f_3$$

$$y_4 = \frac{1}{1+a\,\Delta t}\,y_3 + \frac{\Delta t}{1+a\,\Delta t}\,f_4$$

$$\vdots$$

The backward Euler method is absolutely stable no matter the step size. However, to give sufficient detail in the solution, the step size $\Delta t$ must again be chosen small, resulting in the need for a very large number of iterations.

In this chapter we determine how to obtain *an equation* (the solution to a *difference equation*) that allows us to determine the solution at any $t_n = n\Delta t$ *without having to go through the iterative process* (i.e., *by going directly to the answer*)! The recursion relations above for the two methods can be written in the form of a *difference equation* simply by putting the *y*'s on the left-hand side and the *f*'s on the right. For the forward Euler method, the difference equation becomes

$$y_{n+1} - (1 - a\,\Delta t)y_n = \Delta t f_n$$

and for the backward Euler method the difference equation becomes

$$y_{n+1} - \frac{1}{1+a\,\Delta t}\,y_n = \frac{\Delta t}{1+a\,\Delta t}\,f_{n+1}$$

Using the methods that we develop in this chapter, a *closed-form* result for the *solution* at the *n*th iteration can be obtained for the forward Euler method as

$$y_n = \left(y_0 - \frac{K}{a}\right)(1 - a\,\Delta t)^n + \frac{K}{a}$$

where we assume that $f(t) = K$ is a (known) constant. For the backward Euler method, the solution to the difference equation is

$$y_n = \left(y_0 - \frac{K}{a}\right)\frac{1}{(1+a\,\Delta t)^n} + \frac{K}{a}$$

You can check that these formulas give the same results as the iterative method for the differential equation solved in Section 4.7:

$$\frac{dy(t)}{dt} + \underset{a}{2}\,y(t) = \underset{f(t)}{4} \qquad y(0) = 3$$

After $n = 10$ iterations and for $\Delta t = 0.1$, these closed-form solutions give, for the forward Euler method, $y_{10} = 2.107374$, and for the backward Euler method, $y_{10} = 2.161506$. You can check this by going through the iteration process: You will get precisely these results.

So the advantage of solving the difference equation rather than going through the iterative process using the recursion relations is that *you can go directly to the solution for any n!* Solving the difference equation also gives considerable insight into the behavior of the solution, which is not obtained directly by going through the iterative process. This is another example of the fundamental point that simply using a computer, for example, to solve for the numerical answer does not give significant insight into the behavior of the system that is governed by the equation, which is the fundamental objective of engineering. There are numerous other examples of physical systems that are described by recursion relations and corresponding difference equations. In the remainder of the chapter we determine *closed-form* solutions for these *difference equations*.

## 5.2  HOW TO IDENTIFY LINEAR, CONSTANT-COEFFICIENT, DIFFERENCE EQUATIONS

The general form of a *linear, constant-coefficient, difference equation* is

$$\boxed{y(n+k)+a_1 y(n+k-1)+\cdots+a_{k-1}y(n+1)+a_k y(n)= f(n)} \qquad (5.1)$$

where the index $n$ is the sequence of integers $n = 0, 1, 2, 3, \ldots$ . We henceforth denote

$$y_n \equiv y(n) \quad \text{and} \quad f_n \equiv f(n)$$

Hence, the difference equation can be written in shorthand notation as

$$\boxed{y_{n+k} + a_1 y_{n+k-1} +\cdots+ a_{k-1}y_{n+1} + a_k y_n = f_n} \qquad (5.2)$$

The *order* of the difference equation is the integer $k$. The $k$ coefficients $a_1, a_2,$ $\ldots , a_k$ are known (given), as is the forcing function $f(n) \equiv f_n$ for all values of $n$. Again it is a good idea to make the leading coefficient of the difference equation unity. A difference equation relates the numbers in one sequence, $f(n)$, to the numbers in another sequence, $y(n)$. For example,

$$f_n \equiv f(n) = \left\{ \underset{n=0}{3}, 8, -6, \ldots \right\} \qquad y_n \equiv y(n) = \left\{ \underset{n=0}{2}, -7, -4, \ldots \right\}$$

If we rewrite (5.2) as

$$y_{n+k} = -a_1 y_{n+k-1} - \cdots - a_{k-1} y_{n+1} - a_k y_n + f_n \qquad (5.2)$$

we see that we are obtaining the solution for some value in the sequence $y_{n+k}$, from the (known) values in the sequence $f(n)$, as well as the (known) *previous values* $y_{n+k-1}, \ldots, y_n$. So this is a *recursion relation*.

A difference equation is said to be *linear* if none of the $y$'s are raised to a power greater than unity and/or none of the $y$'s are multiplied together. For example, a second-order, nonlinear, difference equation is

$$y_{n+2} + 3(y_{n+1})^2 + 2y_n y_{n+1} = 5$$

A difference equation is said to be *constant coefficient* if none of the coefficients $a_1, \ldots, a_k$ are functions of the index $n$. An example of a non-constant-coefficient (but linear) third-order difference equation is

$$y_{n+3} + 3y_{n+2} + (2n)y_{n+1} + 6y_n = 10n^2$$

We will study only *linear, constant-coefficient*, difference equations. Linear, constant-coefficient, difference equations are the most common types of difference equations throughout all disciplines of engineering. Nonlinear and/or non-constant-coefficient, difference equations are extremely difficult to solve, so little insight is gained in attempting their solution. They are best solved numerically in an approximate fashion with a digital computer (or your calculator).

Although difference equations of any order occur throughout all the engineering disciplines, the most common are the *first-order difference equation*

$$y_{n+1} + ay_n = f_n$$

and the *second-order difference equation*

$$y_{n+2} + ay_{n+1} + by_n = f_n$$

We investigate only first- and second-order difference equations. In addition, these will be linear and constant coefficient. However, once the solutions to these difference equations are established, you can, on your own, easily extend those ideas to determine the solution to higher-order difference equations.

As was the case for linear, constant-coefficient, ordinary differential equations in Chapter 4, the *total* solution to a linear, constant-coefficient, difference equation is the *sum* of the *homogeneous solution* with $f(n) = 0$,

$$\boxed{y_{n+k}^{(h)} + a_1 y_{n+k-1}^{(h)} + \cdots + a_{k-1} y_{n+1}^{(h)} + a_k y_n^{(h)} = 0} \qquad (5.3a)$$

and the *forced solution* with $f(n) \neq 0$,

$$\boxed{y_{n+k}^{(f)} + a_1 y_{n+k-1}^{(f)} + \cdots + a_{k-1} y_{n+1}^{(f)} + a_k y_n^{(f)} = f_n} \qquad (5.3b)$$

and is

$$\boxed{y_n = y_n^{(h)} + y_n^{(f)}} \qquad (5.3c)$$

It is very helpful to compare the difference equation to the ordinary differential equation of Chapter 4. A comparable $k$th-order linear, constant-coefficient, ordinary differential equation would be

$$\frac{d^k y(t)}{dt^k} + a_1 \frac{d^{(k-1)} y(t)}{dt^{(k-1)}} + \cdots + a_k y(t) = f(t)$$

By comparing this to the general form of a $k$th-order difference equation in (5.2), we can make an analogy between the difference equation and the differential equation. We will find that the subjects of Chapter 4 (and the outline of that chapter) will follow those of this chapter in almost identical fashion. For example, we investigate only first- and second-order difference equations where $k = 1$ and $k = 2$. The total solution will again be the sum of a homogeneous solution with $f(n) = 0$ and a forced solution with $f(n) \neq 0$. The forced solution can also be found by the "method of undetermined coefficients." The reader should keep this analogy in mind, as it will be helpful in at least anticipating what to do next.

## 5.3  SOLUTION OF FIRST-ORDER EQUATIONS

The first-order difference equation is

$$\boxed{y_{n+1} + a y_n = f_n} \qquad (5.4a)$$

The *homogeneous solution* is the solution with $f_n = 0$:

$$\boxed{y_{n+1}^{(h)} + a y_n^{(h)} = 0} \qquad (5.4b)$$

and the *forced solution* is the solution with $f_n \neq 0$:

$$\boxed{y_{n+1}^{(f)} + a y_n^{(f)} = f_n} \qquad (5.4c)$$

The *total solution* is the sum of these:

$$\boxed{y_n = y_n^{(h)} + y_n^{(f)}} \qquad (5.4d)$$

In addition, we must specify the *initial condition* for the particular problem as $y_0 \equiv y(0)$.

### 5.3.1   The Homogeneous Solution

The homogeneous solution satisfies (5.4b):

$$\boxed{y_{n+1}^{(h)} + ay_n^{(h)} = 0} \tag{5.4b}$$

To satisfy this, the homogeneous solution must have the property that $y_n^{(h)}$ and *its value shifted*, $y_{n+1}^{(h)}$, must have the same *form*. The function $r^n$ has the property that $r^{(n+k)} = r^n r^k$. So we choose the homogeneous solution to be of the *form*

$$\boxed{y_n^{(h)} = Cr^n} \tag{5.5}$$

where $r$ is to be determined. Substituting this into (5.4b) yields

$$Cr^{(n+1)} + aCr^n = Cr^n r + aCr^n$$
$$= (r+a)Cr^n$$
$$= 0$$

The only way this can be satisfied is to satisfy the *characteristic equation*:

$$r + a = 0$$

which gives the value of $r$ as

$$r = -a$$

Hence, the homogeneous solution is

$$\boxed{y_n^{(h)} = C(-a)^n} \tag{5.6}$$

and $C$ is, as yet, an undetermined constant. It will be determined when we apply the initial condition $y_0 \equiv y(0)$ to the *total solution*.

### 5.3.2   The Forced Solution for "Nice" $f(n)$

As in the case of ordinary differential equations in Chapter 4, we determine the forced solution using the "method of undetermined coefficients," where we assume a *form* of $y_n^{(f)}$ that is of the same *form* as $f(n)$ but with some undetermined constants in it and substitute into

$$y_{n+1}^{(f)} + a y_n^{(f)} = f_n \tag{5.4c}$$

to determine those constants. There are three very common forms of $f(n)$ that are frequently found in engineering applications.

### $f(n)$ Is the $n$th Power of a Constant

$$\boxed{f(n) = K\alpha^n} \tag{5.7a}$$

where $K$ and $\alpha$ are known (given). We assume the forced solution to be of the same *form*:

$$\boxed{y_n^{(f)} = M\alpha^n} \tag{5.7b}$$

and substitute into (5.4c) to determine $M$:

$$M\alpha^{(n+1)} + aM\alpha^n = K\alpha^n$$

After factoring and canceling the term $\alpha^n$ that is common to both sides, we obtain

$$\boxed{M = \frac{K}{\alpha + a}} \tag{5.7c}$$

### $f(n)$ Is a Polynomial in $n$

$$\boxed{f(n) = \alpha^n \left( K_0 + K_1 n + K_2 n^2 + \cdots + K_m n^m \right)} \tag{5.8a}$$

We assume the forced solution to be a similar form:

$$\boxed{y_n^{(f)} = \alpha^n \left( M_0 + M_1 n + M_2 n^2 + \cdots + M_m n^m \right)} \tag{5.8b}$$

and substitute into (5.4c). Comparing corresponding coefficients of powers of $n$ on both sides gives equations to solve for the $m + 1$ unknowns $M_0, M_1, \dots, M_m$. Observe, as was the case for differential equations in Chapter 4, that we must use all the $M_i$ coefficients of the lower powers of $n$ in the assumed form for $y_n^{(f)}$, even though some of the coefficients of the lower powers of $n$ are absent in $f(n)$. For example, if $f(n) = 3n^2$, we must assume the form $y_n^{(f)} = M_0 + M_1 n + M_2 n^2$.

### $f(n)$ Is a Sinusoid

$$\boxed{f(n) = K\alpha^n \cos n\theta \quad \text{or} \quad K\alpha^n \sin n\theta} \tag{5.9}$$

where $K$, $\alpha$, and $\theta$ are known (given). We use the same technique that was used in Chapter 4 to determine the forced solution for sinusoidal forcing functions. Using Euler's identity for complex numbers, we replace $f(n)$ with

$$f(n) \Rightarrow K\alpha^n e^{jn\theta} \tag{5.10a}$$

and the forced solution for this forcing function now satisfies

$$\mathbf{Y}_{n+1}^{(f)} + a\mathbf{Y}_n^{(f)} = K\alpha^n e^{jn\theta} \tag{5.10b}$$

By the method of undetermined coefficients, we assume this forced solution to be of the same form as this $f(n)$:

$$\mathbf{Y}_n^{(f)} = \mathbf{M}\alpha^n e^{jn\theta} \tag{5.10c}$$

where $\mathbf{M}$ is now complex. Substituting (5.10c) into (5.10b) to determine $\mathbf{M}$ gives

$$\mathbf{M}\alpha^{(n+1)} e^{j(n+1)\theta} + a\mathbf{M}\alpha^n e^{jn\theta} = K\alpha^n e^{jn\theta}$$

Factoring this gives

$$(\alpha e^{j\theta} + a)\mathbf{M}\alpha^n e^{jn\theta} = K\alpha^n e^{jn\theta}$$

Canceling common terms on both sides gives

$$\boxed{\begin{aligned} \mathbf{M} &= \frac{K}{\alpha e^{j\theta} + a} \\ &= M\angle\phi \end{aligned}} \tag{5.10d}$$

(Note that $e^{j\theta} = 1\angle\theta$.) Hence, using Euler's identity, the forced solution is

$$\boxed{y_n^{(f)} = M\alpha^n \cos(n\theta + \phi) \quad \text{or} \quad M\alpha^n \sin(n\theta + \phi)} \tag{5.10e}$$

depending on whether the original $f(n)$ was a cosine or a sine.


### 5.3.3  The Total Solution

The total solution is the sum of the homogeneous and forced solutions:

$$\begin{aligned} y_n &= y_n^{(h)} + y_n^{(f)} \\ &= C(-a)^n + y_n^{(f)} \end{aligned}$$

The final task is to apply the initial condition to this total solution in order to evaluate the undetermined constant $C$:

$$y_0 = C + y_0^{(f)} \tag{5.11}$$

Solving (5.11) for $C$ gives the *total solution* as

$$\boxed{y_n = [y_0 - y_0^{(f)}](-a)^n + y_n^{(f)}} \tag{5.12}$$

---

### Example

Determine the total solution to the following difference equation subject to the initial condition $y_0 = 3$:

$$y_{n+1} + 2y_n = \underbrace{4(-3)^n}_{f_1(n)} + \underbrace{2n}_{f_2(n)} + \underbrace{6\cos(n\frac{\pi}{2})}_{f_3(n)} \qquad y_0 = 3$$

The characteristic equation is

$$r + 2 = 0$$

so that $r = -2$, giving the homogeneous solution as

$$y_n^{(h)} = C(-2)^n$$

The forced solution due to $f_1(n) = 4(-3)^n$ is assumed to be of the form $M(-3)^n$. Substituting gives

$$y_{n+1}^{(f)} + 2y_n^{(f)} = M(-3)^{(n+1)} + 2M(-3)^n = 4(-3)^n$$

or

$$M = \frac{4}{(-3)+2} = -4$$

The forced solution due to $f_2(n) = 2n$ is assumed to be of the form $M_1 + M_2n$. Note again that you must use all lower powers of $n$ in the assumed form of the forced solution if $f(n)$ contains a power of $n$. Substituting gives

$$y_{n+1}^{(f)} + 2y_n^{(f)} = M_1 + M_2(n+1) + 2M_1 + 2M_2n = 2n$$

Matching coefficients of corresponding powers of $n$ on both sides gives

$$3M_1 + M_2 = 0$$
$$3M_2 = 2$$

Solving these two simultaneous equations gives

$$M_1 = -\frac{2}{9}, \qquad M_2 = \frac{2}{3}$$

The forced solution due to

$$f_3(n) = 6\cos\left(n\frac{\pi}{2}\right)$$

is obtained by replacing

$$f_3(n) = 6\cos\left(n\frac{\pi}{2}\right) \Rightarrow 6e^{jn(\pi/2)}$$

The form of the forced solution is assumed to be $\mathbf{M}e^{jn(\pi/2)}$. Substituting gives

$$\mathbf{Y}_{n+1}^{(f)} + 2\mathbf{Y}_n^{(f)} = \mathbf{M}e^{j(n+1)(\pi/2)} + 2\mathbf{M}e^{jn(\pi/2)} = 6e^{jn(\pi/2)}$$

Canceling terms common to both sides gives

$$\mathbf{M} = \frac{6}{\underbrace{e^{j(\pi/2)}}_{j} + 2} = \frac{6}{j+2} = 2.68\angle - 26.57°$$

Hence, the total forced solution is the sum of these:

$$y_n^{(f)} = -4(-3)^n - \frac{2}{9} + \frac{2}{3}n + 2.68\cos\left(n\frac{\pi}{2} - 26.57°\right)$$

Thus, the total solution is

$$y_n = C(-2)^n - 4(-3)^n - \frac{2}{9} + \frac{2}{3}n + 2.68\cos\left(n\frac{\pi}{2} - 26.57°\right)$$

Apply the initial condition, $y_0 = 3$, to this total solution to give $C = 4.82$. (You should verify this.)

### 5.3.4  A Special Case

> *If the forcing function f(n) has the same form as in the homogeneous solution, the usual assumed form of the forced solution must be multiplied by n.*

The reason for this is that if the forcing function is $f_n = K(-a)^n$, it will have the same form as the homogeneous solution $y_n^{(h)} = C(-a)^n$. But the homogeneous solution satisfies the *homogeneous* difference equation in (5.4b), whose right-hand side is zero. Therefore, choosing the usual form of the forced solution as $y_n^{(f)} = M(-a)^n$ cannot satisfy the nonhomogeneous difference equation in (5.4c), whose right-hand side is not zero. But multiplying the usual assumed form by $n$ to give $y_n^{(f)} = nM(-a)^n$ will satisfy (5.4c), as the following example shows.

---

### *Example*

Determine the form of the total solution to the difference equation

$$y_{n+1} + 3y_n = 2(-3)^n$$

The homogeneous solution is

$$y_n^{(h)} = C(-3)^n$$

Since the homogeneous solution has the same form as the $f(n)$, we multiply the usual assumed form of the forced solution by $n$ and assume that

$$y_n^{(f)} = Mn(-3)^n$$

Substituting gives

$$M(n+1)(-3)^{(n+1)} + 3Mn(-3)^n = 2(-3)^n$$

Expanding this and noting that $(-3)^{(n+1)} = (-3)(-3)^n$ gives

$$Mn(-3)(-3)^n + M(-3)(-3)^n + 3Mn(-3)^n = 2(-3)^n$$

Canceling common terms on both sides gives

$$M = -\frac{2}{3}$$

Hence, the forced solution is

$$y_n^{(f)} = -\frac{2}{3}n(-3)^n$$

and the total solution is

$$y_n = C(-3)^n - \frac{2}{3}n(-3)^n$$

Evaluating this for the initial condition $y_0$ gives the total solution

$$y_n = y_0(-3)^n - \frac{2}{3}n(-3)^n$$

which you should verify.

---

## 5.4   SOLUTION OF SECOND-ORDER EQUATIONS

Finally, we investigate the solution to the second-order difference equation subject to the two initial conditions $y_0$ and $y_1$:

$$\boxed{y_{n+2} + ay_{n+1} + by_n = f_n} \qquad \text{for } y_0, y_1 \qquad (5.13)$$

Again, the total solution is the sum of the homogeneous and the forced solutions because the difference equation is *linear*.

### 5.4.1   The Homogeneous Solution

The *homogeneous* solution satisfies

$$\boxed{y_{n+2}^{(h)} + ay_{n+1}^{(h)} + by_n^{(h)} = 0} \qquad (5.14)$$

Again we assume the form of the homogeneous solution to be

$$y_n^{(h)} = Cr^n$$

and need to determine $r$. Substituting this into (5.14) gives

$$(r^2 + ar + b)Cr^n = 0$$

This can only be satisfied by determining the two values of $r$ that satisfy the second-order *characteristic equation*

$$\boxed{r^2 + ar + b = (r - r_1)(r - r_2) = 0} \tag{5.15}$$

As was the case for ordinary differential equations in Chapter 4, the two roots can be (a) real and distinct, (b) real and repeated, or (c) complex conjugate. The forms of the homogeneous solution for each of these are very similar to those for ordinary differential equations.

### Roots Real and Distinct: $r_1 \neq r_2$

$$\boxed{y_n^{(h)} = C_1 r_1^n + C_2 r_2^n} \tag{5.16}$$

The two undetermined constants $C_1$ and $C_2$ will be determined by applying the two initial conditions $y_0$ and $y_1$ to the *total solution* once we have determined the forced solution.

### Roots Real and Repeated: $r_1 = r_2 = r$

In this case of identical roots, the two terms in (5.16) could be combined, thereby eliminating one of the undetermined constants. But we have two initial conditions to be applied to determine these undetermined constants. So for repeated roots we multiply either the first or the second terms in (5.16) by $n$ and assume the form

$$\boxed{y_n^{(h)} = C_1 r^n + C_2 n r^n} \tag{5.17}$$

We can show that the second piece of the homogeneous solution, $C_2 n r^n$, also satisfies the homogeneous differential equation for repeated roots. For repeated roots the characteristic equation becomes $r^2 + ar + b = (r - \hat{r})(r - \hat{r}) = r^2 - 2\hat{r}r + \hat{r}^2 = 0$. We have denoted the two repeated roots as $r = \hat{r}$ here to avoid confusion. We identify $a = -2\hat{r}$ and $b = \hat{r}^2$. Substituting $C_2 n \hat{r}^n$ into the homogeneous equation in (5.14) gives

$$C_2(n+2)\hat{r}^{(n+2)} - 2\hat{r}C_2(n+1)\hat{r}^{(n+1)} + \hat{r}^2 C_2 n \hat{r}^n \overset{?}{=} 0$$

Writing this out gives

$$[(n\hat{r}^2 + 2\hat{r}^2) - 2\hat{r}(n\hat{r} + \hat{r}) + \hat{r}^2 n] C_2 \hat{r}^n \overset{?}{=} 0$$

Canceling terms, we see that this is satisfied.

What would you choose for the form of the homogeneous solution in the case of a third-order difference equation that has all three roots identical: $r_1 = r_2 = r_3 = r$?

***Roots Complex Conjugate:*** $r_1 = \alpha + j\beta$, $r_2 = \alpha - j\beta$   Since these are distinct roots the homogeneous solution is the same as (5.16):

$$\boxed{y_n^{(h)} = C_1(\alpha + j\beta)^n + C_2(\alpha - j\beta)^n} \tag{5.18a}$$

We can also generate equivalent forms using Euler's identity. To do so, write the roots in *polar form* as

$$\boxed{(\alpha + j\beta) = \rho \angle \theta \quad \text{and} \quad (\alpha - j\beta) = \rho \angle -\theta} \tag{5.18b}$$

where

$$\rho = \sqrt{\alpha^2 + \beta^2} \quad \text{and} \quad \theta = \tan^{-1}\frac{\beta}{\alpha} \tag{5.18c}$$

Remember that $\rho \angle \theta \equiv \rho e^{j\theta}$. Using Euler's identity, we can generate the following equivalent forms:

$$\boxed{y_n^{(h)} = K_1 \rho^n e^{jn\theta} + K_2 \rho^n e^{-jn\theta}} \tag{5.18d}$$

or

$$\boxed{y_n^{(h)} = M_1 \rho^n \cos n\theta + M_2 \rho^n \sin n\theta} \tag{5.18e}$$

or

$$\boxed{y_n^{(h)} = N_1 \rho^n \cos(n\theta + \phi_1)} \tag{5.18f}$$

or

$$\boxed{y_n^{(h)} = N_2 \rho^n \sin(n\theta + \phi_2)} \tag{5.18g}$$

Applying the two initial conditions $y_0$ and $y_1$ to the *total solution* allows the determination of the two undetermined constants: $K_1$ and $K_2$, or $M_1$ and $M_2$, or $N_1$ and $\phi_1$, or $N_2$ and $\phi_2$, depending on the form you choose to use.

These equivalent forms are obtained using Euler's identity from Chapter 2:

$$e^{jn\theta} = 1 \angle n\theta = \cos n\theta + j \sin n\theta$$

For example, (5.18e) can be obtained from (5.18d) by substituting Euler's identity to give

$$y_n^{(h)} = K_1\rho^n e^{jn\theta} + K_2\rho^n e^{-jn\theta}$$
$$= K_1\rho^n(\cos n\theta + j\sin n\theta) + K_2\rho^n(\cos n\theta - j\sin n\theta)$$
$$= \underbrace{(K_1 + K_2)}_{M_1}\rho^n \cos n\theta + \underbrace{j(K_1 - K_2)}_{M_2}\rho^n \sin n\theta \qquad (5.18d)$$

Since $y_n^{(h)}$ must be a real number (for physical systems), $M_1$ and $M_2$ must be real numbers. For this to be the case, the undetermined constants $K_1$ and $K_2$ in the original form in (5.18d) must be complex numbers and must be the conjugates of each other: $K_2 = K_1^*$, where * denotes the conjugate of a complex number. For example, if we write $K_1 = a + jb$ and $K_2 = K_1^* = a - jb$, then $M_1 = K_1 + K_2 = K_1 + K_1^* = 2a$ and $M_2 = j(K_1 - K_2) = j(K_1 - K_1^*) = -2b$, which are both real numbers. Hence, $y_n^{(h)}$ in (5.18d) and (5.18e) is a real number, as it should be in order to be the response of a physical system.

Similarly, (5.18f) can be obtained from (5.18d) by writing the undetermined coefficients in polar form as $K_1 = N\angle\phi = Ne^{j\phi}$ and $K_2 = K_1^* = N\angle-\phi = Ne^{-j\phi}$. Substituting gives

$$y_n^{(h)} = K_1\rho^n e^{jn\theta} + K_2\rho^n e^{-jn\theta}$$
$$= K_1\rho^n e^{jn\theta} + K_1^*\rho^n e^{-jn\theta}$$
$$= Ne^{j\phi}\rho^n e^{jn\theta} + Ne^{-j\phi}\rho^n e^{-jn\theta}$$
$$= N\rho^n(e^{j(n\theta+\phi)} + e^{-j(n\theta+\phi)})$$
$$= 2N\rho^n \cos(n\theta + \phi) \qquad (5.18d)$$

which is the form in (5.18f). You should similarly verify the form in (5.18g). *Hint*: Write $K_1 = jN\angle\phi = jNe^{j\phi}$ and $K_2 = K_1^* = -jN\angle-\phi = -jNe^{-j\phi}$.

### 5.4.2  The Forced Solution for "Nice" $f(n)$

The forced solution $y_n^{(f)}$ for second-order equations is the solution to

$$\boxed{y_{n+2}^{(f)} + ay_{n+1}^{(f)} + by_n^{(f)} = f_n} \qquad (5.19)$$

It is obtained in the same fashion as for first-order equations: Use the method of undetermined coefficients.

### 5.4.3  The Total Solution

The total solution is formed by adding the homogeneous and forced solutions:

$$\boxed{y_n = y_n^{(h)} + y_n^{(f)}} \qquad (5.20)$$

We evaluate the two undetermined constants in the homogeneous solution by applying the two initial conditions, $y_0$ and $y_1$, to this total solution.

---

## *Example*

Determine the solution to the following difference equation:

$$y_{n+2} + 6y_{n+1} + 8y_n = \underset{f(n)}{5}$$

The characteristic equation is

$$r^2 + 6r + 8 = (r+2)(r+4) = 0$$

whose roots are

$$r_1 = -2 \quad \text{and} \quad r_2 = -4$$

The form of the homogeneous solution is

$$y_n^{(h)} = C_1(-2)^n + C_2(-4)^n$$

The form of the forced solution is $y_n^{(f)} = M$. Substituting yields

$$M + 6M + 8M = 5$$

Solving gives $M = 1/3$, so that the forced solution is

$$y_n^{(f)} = \frac{1}{3}$$

Hence, the total solution is

$$y_n = C_1(-2)^n + C_2(-4)^n + \frac{1}{3}$$

You should "sanity check" this result. Applying the initial conditions, $y_0$ and $y_1$, gives two equations that can be solved by the methods of Chapter 3 for the undetermined constants, $C_1$ and $C_2$:

$$y_0 = C_1 + C_2 + \frac{1}{3}$$

$$y_1 = -2C_1 - 4C_2 + \frac{1}{3}$$

which you should verify.

### *Example*

Determine the solution to the following difference equation:

$$y_{n+2} + 6y_{n+1} + 9y_n = \underbrace{5(-2)^n}_{f(n)}$$

The characteristic equation is

$$r^2 + 6r + 9 = (r+3)(r+3) = 0$$

whose roots are

$$r_1 = r_2 = -3$$

The form of the homogeneous solution is

$$y_n^{(h)} = C_1(-3)^n + C_2 n(-3)^n$$

The form of the forced solution is $y_n^{(f)} = M(-2)^n$. Substituting yields

$$M(-2)^{(n+2)} + 6M(-2)^{(n+1)} + 9M(-2)^n = 5(-2)^n$$

Expanding gives

$$M(-2)^n(-2)^2 + 6M(-2)^n(-2) + 9M(-2)^n = 5(-2)^n$$

Canceling the $(-2)^n$ on both sides gives $M = 5$, so that

$$y_n^{(f)} = 5(-2)^n$$

Hence, the total solution is

$$y_n = C_1(-3)^n + C_2 n(-3)^n + 5(-2)^n$$

You should "sanity check" this result. Applying the initial conditions, $y_0$ and $y_1$, gives two equations that can be solved by the methods of Chapter 3 for the undetermined constants, $C_1$ and $C_2$:

$$y_0 = C_1 + 5$$
$$y_1 = -3C_1 - 3C_2 - 10$$

which you should verify.

## *Example*

Determine the solution to the following difference equation:

$$y_{n+2} + 4y_{n+1} + 13y_n = \underbrace{5n}_{f(n)}$$

The characteristic equation is

$$r^2 + 4r + 13 = (r + 2 - j3)(r + 2 + j3) = 0$$

whose roots are

$$r_1 = -2 + j3 = 3.61\angle 123.69°$$
$$r_2 = -2 - j3 = 3.61\angle -123.69°$$

The form (one of several possible for these complex roots) of the homogeneous solution is

$$y_n^{(h)} = C_1 (3.61)^n \cos(n \cdot 123.69°) + C_2 (3.61)^n \sin(n \cdot 123.69°)$$

The form of the forced solution is $y_n^{(f)} = M_1 + M_2 n$. Substituting yields

$$M_1 + M_2(n+2) + 4M_1 + 4M_2(n+1) + 13M_1 + 13M_2 n = 5n$$

Comparing corresponding coefficients of powers of $n$ on both sides gives two equations that can be solved by the methods of Chapter 3 for $M_1$ and $M_2$:

$$18M_1 + 6M_2 = 0$$
$$18M_2 = 5$$

Solving these gives $M_1 = -5/54$ and $M_2 = 5/18$ (which you should "sanity check"), so that the forced solution is

$$y_n^{(f)} = -\frac{5}{54} + \frac{5}{18}n$$

Hence, the total solution is the sum of the homogeneous and forced solutions:

$$y_n = C_1 (3.61)^n \cos(n \cdot 123.69°) + C_2 (3.61)^n \sin(n \cdot 123.69°) - \frac{5}{54} + \frac{5}{18}n$$

Applying the initial conditions, $y_0$ and $y_1$, gives two equations that can be solved by the methods of Chapter 3 for the undetermined constants, $C_1$ and $C_2$:

$$y_0 = C_1 - \frac{5}{54}$$

$$y_1 = 3.61\cos(123.69°)C_1 + 3.61\sin(123.69°)C_2 - \frac{5}{54} + \frac{5}{18}$$

which you should verify.

---

### 5.4.4  A Special Case

If the forcing function $f(n)$ has the same *form* as any part of the *homogeneous solution*, we cannot use the usual assumed form of the forced solution. The reason is that this would satisfy the homogeneous equation whose right-hand side is *zero*. Hence, it could not satisfy the nonhomogeneous equation whose right-hand side is $f(n)$, which is *not* zero. In this case the usual assumed form of the forced solution must be multiplied by $n$ (distinct roots) or $n^2$ (repeated roots).

*Distinct Roots (Which Includes Complex Roots): $r_1 \neq r_2$*

$$\boxed{\begin{array}{c} f(n) = Kr_1^n \Rightarrow y_n^{(f)} = Mnr_1^n \\ \text{or} \\ f(n) = Kr_2^n \Rightarrow y_n^{(f)} = Mnr_2^n \end{array}} \qquad (5.21)$$

---

*Example*

Determine the solution to the following difference equation:

$$y_{n+2} + 3y_{n+1} + 2y_n = 5(-2)^n$$

The characteristic equation is

$$r^2 + 3r + 2 = (r+1)(r+2) = 0$$

whose roots are $r_1 = -1$ and $r_1 = -2$. Hence, the homogeneous solution is

$$y_n^{(h)} = C_1(-1)^n + C_2(-2)^n$$

We cannot assume that $y_n^{(f)} = M(-2)^n$ because that would be the same form as $C_2(-2)^n$ in the homogeneous solution which satisfies the homogeneous equation whose right-hand side is *zero*. It could therefore not be made to satisfy the nonhomogeneous equation whose right-hand side is *not* zero. But $y_n^{(f)} = Mn(-2)^n$ would work:

$$
\begin{aligned}
y_{n+2}^{(f)} + 3y_{n+1}^{(f)} + 2y_n^{(f)} &= M(n+2)(-2)^{(n+2)} + 3M(n+1)(-2)^{(n+1)} + 2Mn(-2)^n \\
&= M\left[n(-2)^2 + 2(-2)^2 + 3n(-2) + 3(-2) + 2n\right](-2)^n \\
&= M\left[2(-2)^2 + 3(-2)\right](-2)^n \\
&= 2M(-2)^n \\
&= 5(-2)^n
\end{aligned}
$$

Hence, $M = 5/2$ and $y_n^{(f)} = (5/2)n(-2)^n$.

---

***Repeated Roots:*** $r_1 = r_2 = r$

$$\boxed{f(n) = Kr^n \Rightarrow y_n^{(f)} = Mn^2 r^n} \qquad (5.22)$$

---

### *Example*

Determine the solution to the following difference equation:

$$y_{n+2} + 4y_{n+1} + 4y_n = 5(-2)^n$$

The characteristic equation is

$$r^2 + 4r + 4 = (r+2)(r+2) = 0$$

Hence, the two roots are repeated, $r_1 = r_2 = -2$, and the homogeneous solution is

$$y_n^{(h)} = C_1(-2)^n + C_2 n(-2)^n$$

We cannot assume that $y_n^{(f)} = M(-2)^n$ because that would be the same form as $C_1(-2)^n$ in the homogeneous solution that satisfies the homogeneous equation whose right-hand side is *zero*. It could therefore not be made to satisfy the nonhomogeneous equation whose right-hand side is *not* zero. We could also not assume that $y_n^{(f)} = Mn(-2)^n$ because that would be the same form as $C_2 n(-2)^n$ in the homogeneous solution that satisfies the homogeneous equation whose right-hand side is *zero*. It could therefore not be made to satisfy the nonhomogeneous equation whose right-hand side is *not* zero. But $y_n^{(f)} = Mn^2(-2)^n$ would work:

$$
\begin{aligned}
y_{n+2}^{(f)} + 4y_{n+1}^{(f)} + 4y_n^{(f)} &= M(n+2)^2 (-2)^{(n+2)} + 4M(n+1)^2 (-2)^{(n+1)} + 4Mn^2(-2)^n \\
&= M\left[(n^2 + 4n + 4)(-2)^2 + 4(n^2 + 2n + 1)(-2) + 4n^2\right](-2)^n \\
&= 8M(-2)^n \\
&= 5(-2)^n
\end{aligned}
$$

Hence, $M = 5/8$ and $y_n^{(f)} = (5/8)n^2 (-2)^n$.

---

## 5.5   STABILITY OF THE SOLUTION

The homogeneous solution should go to zero as $n \to \infty$ for physical systems, thereby giving the forced solution as the total solution. If not, we say that the solution is *unstable*. The homogeneous solution is of the form

$$
y_n^{(h)} = C_1(r_1)^n + C_2(r_2)^n
$$

Since the homogeneous solution involves the *n*th powers of the roots of the characteristic equation, these terms will decrease to zero as $n \to \infty$ only if those roots have absolute values less than unity (i.e., $|r_i| < 1$).

In the numerical solution of a first-order ordinary differential equation by the forward Euler method in Chapter 4 we obtained a recursion relation or difference equation

$$
y_{n+1} - (1 - a\,\Delta t)y_n = \Delta t\, f(n)
$$

where the root of the characteristic equation is

$$
r = 1 - a\,\Delta t
$$

Hence, for this solution method to be stable, $|r| = |1 - a\Delta t| < 1$, we must have $\Delta t < 2/a$. For the example in Chapter 4 where $a = 2$, we found that instability of the solution occurs when we choose $\Delta t \geq 1$.

## 5.6   SOLUTION OF SIMULTANEOUS SETS OF DIFFERENCE EQUATIONS WITH THE DIFFERENCE OPERATOR

The difference operator $D$ "operates on" a sequence, $f_n$, to shift that sequence:

$$\boxed{D^m f_n \Rightarrow f_{n+m}}$$ 
(5.23)

The difference operator obeys the rules of algebra and can be treated like an algebraic quantity:

$$D^m(f_n + g_n) = f_{n+m} + g_{n+m}$$

$$D^k D^m f_n = D^{k+m} f_n = f_{n+k+m}$$

$$\frac{1}{D^k} D^m f_n = D^{m-k} f_n = f_{n+m-k}$$

$$\frac{1}{D^m} f_n = f_{n-m}$$

The $D$ operator *must appear on the left of the function* or it wouldn't make sense. For example, writing $f_n D$ wouldn't make sense because there is no function on the right of $D$ for it to "operate on."

The difference operator can be used to write difference equations in a tidy, symbolic form. One of the primary advantages of doing this with the difference operator is that *simultaneous sets* of difference equations can be manipulated and solved easily, as the following examples show.

---

### *Example*

Write the following set of two simultaneous difference equations in symbolic form using the difference operator:

$$3x_{n+2} + 2x_{n+1} + 5y_{n+1} + y_n = 2^n$$

$$x_{n+1} - 3x_n + 2y_{n+2} = 5n^2$$

Using the difference operator gives the symbolic differential equations

$$(3D^2 + 2D)x_n + (5D + 1)y_n = 2^n$$

$$(D - 3)x_n + (2D^2)y_n = 5n^2$$

which you should verify.

---

The problem with simultaneous sets of difference equations is that each difference equation in the set involves more than one variable. For example, each equation in the simultaneous difference equations of the preceding example contains *both* $x_n$ and $y_n$. The two equations in the set are therefore said to be *coupled*. To solve the simultaneous set, we must be able to reduce them to two *uncoupled* difference equations, each involving *only* one variable, such as $x_n$ or $y_n$. The difference operator reduces the coupled set of *difference equations* to a coupled set of *algebraic equations* that we know how to solve (Chapter 3). These can then be manipulated to obtain uncoupled difference equations. The following example illustrates this valuable property of the difference operator.

---

## *Example*

Write the following *coupled* difference equations using the difference operator as in the preceding example:

$$x_{n+1} + x_n + y_n = 2n$$

$$-2x_n + y_{n+1} + 4y_n = 5n^2$$

$$\Downarrow$$

$$(D+1)x_n + y_n = 2n$$

$$-2x_n + (D+4)y_n = 5n^2$$

Reduce them to two difference equations, each of which contains only one unknown, $x_n$ or $y_n$, but not both. To do so we may write these in matrix form (Chapter 3) as

$$\begin{bmatrix} D+1 & 1 \\ -2 & D+4 \end{bmatrix} \begin{bmatrix} x_n \\ y_n \end{bmatrix} = \begin{bmatrix} 2n \\ 5n^2 \end{bmatrix}$$

Solving by Cramer's rule gives

$$x_n = \frac{\begin{vmatrix} 2n & 1 \\ 5n^2 & D+4 \end{vmatrix}}{\begin{vmatrix} D+1 & 1 \\ -2 & D+4 \end{vmatrix}} = \frac{(D+4)2n - 5n^2}{D^2 + 5D + 6} = \frac{2(n+1) + 8n - 5n^2}{D^2 + 5D + 6}$$

and

$$y_n = \frac{\begin{vmatrix} D+1 & 2n \\ -2 & 5n^2 \end{vmatrix}}{\begin{vmatrix} D+1 & 1 \\ -2 & D+4 \end{vmatrix}} = \frac{(D+1)5n^2 + 4n}{D^2 + 5D + 6} = \frac{5(n+1)^2 + 5n^2 + 4n}{D^2 + 5D + 6}$$

Multiplying both sides by the denominator gives the individual difference equations as

$$(D^2 + 5D + 6)x_n = 2(n+1) + 8n - 5n^2$$
$$(D^2 + 5D + 6)y_n = 5(n+1)^2 + 5n^2 + 4n$$

or

$$x_{n+2} + 5x_{n+1} + 6x_n = 2 + 10n - 5n^2$$
$$y_{n+2} + 5y_{n+1} + 6y_n = 5 + 14n + 10n^2$$

These individual difference equations can be solved by the previous methods, giving the total solutions as

$$x_n = A_1(-2)^n + A_2(-3)^n - 0.291 + 1.319n - 0.417n^2$$
$$y_n = B_1(-2)^n + B_2(-3)^n - 0.321 + 0.194n + 0.833n^2$$

which you should verify.

---

The four undetermined constants in the homogeneous solutions for the coupled difference equations in the preceding example are related.

---

### Example

For example, write the first difference equation,

$$x_{n+1} + x_n + y_n = 2n$$

as

$$y_n = -x_{n+1} - x_n + 2n$$

Write the solution for $x_n$ as

$$x_n = \underbrace{A_1(-2)^n + A_2(-3)^n}_{x_n^{(h)}} + x_n^{(f)}$$

Substitute to give

$$
\begin{aligned}
y_n &= -x_{n+1} - x_n + 2n \\
&= -A_1(-2)^{n+1} - A_2(-3)^{n+1} - A_1(-2)^n - A_2(-3)^n - x_{n+1}^{(f)} - x_n^{(f)} + 2n \\
&= A_1(-2)^n + 2A_2(-3)^n - x_{n+1}^{(f)} - x_n^{(f)} + 2n
\end{aligned}
$$

Comparing this to the solution for $y_n$,

$$y_n = \underbrace{B_1(-2)^n + B_2(-3)^n}_{y_n^{(h)}} + y_n^{(f)}$$

we see that the undetermined coefficients in the homogeneous solution for $y_n$ are related to those in the homogeneous solution for $x_n$ as

$$B_1 = A_1$$
$$B_2 = 2A_2$$

Hence, *the number of initial conditions that can be specified is equal to the order of the determinant in the denominator of Cramer's rule:* $D^2 + 5D + 6$. We could specify

$$x(0) \quad \text{and} \quad x(1)$$

or

$$y(0) \quad \text{and} \quad y(1)$$

or

$$x(0) \quad \text{and} \quad y(1)$$

or

$$y(0) \quad \text{and} \quad x(1)$$

---

### 5.6.1 Using the Difference Operator to Verify Solutions

The difference operator can also be used *easily* and *immediately* to verify the various forms of the homogeneous and forced solutions. First write the second-order difference equation

$$y_{n+2} + ay_{n+1} + by_n = f_n$$

using the difference operator as

$$(D^2 + aD + b)y_n = f_n$$

Write this in factored form as

$$(D - r_1)(D - r_2)y_n = f_n \tag{5.24}$$

where $r_1$ and $r_2$ are the roots of the characteristic equation. The important functions that appear in the various solutions are $r^n$, $nr^n$, $n^2r^n$, and so on. First we establish some important results:

$$(D - r)r^n = r^{(n+1)} - r^{(n+1)} = 0 \tag{5.25a}$$

$$(D - r)nr^n = (n+1)r^{(n+1)} - nr^{(n+1)} = r^{(n+1)} \tag{5.25b}$$

$$(D - r)n^2r^n = (n+1)^2 r^{(n+1)} - n^2 r^{(n+1)} = (2n+1)r^{(n+1)} \tag{5.25c}$$

$$(D - r)n^3r^n = (n+1)^3 r^{(n+1)} - n^3 r^{(n+1)} = (3n^2 + 3n + 1)r^{(n+1)} \tag{5.25d}$$

Using these fundamental results we are now able to verify easily that the previously determined forms of the homogeneous and forced solutions indeed satisfy the appropriate differential equation.

***Homogeneous Solution*** $f_n = 0$, ***Distinct Roots:*** $r_1 \neq r_2$    The homogeneous solution for this case is the sum of the two functions $C_1 r_1^n$ and $C_2 r_2^n$. Substituting $C_1 r_1^n$ into (5.24) gives

$$(D - r_1)(D - r_2)C_1 r_1^n \overset{?}{=} 0$$

Interchanging the operators and using (5.25a) shows immediately that

$$(D - r_2)\underbrace{(D - r_1)C_1 r_1^n}_{0} \overset{?}{=} 0$$

Similarly, substituting $C_2 r_2^n$ into (5.24) and using (5.25a) shows immediately that

$$(D - r_1)\underbrace{(D - r_2)C_2 r_2^n}_{0} \overset{?}{=} 0$$

So the difference operator is a very simple means of demonstrating immediately that these assumed forms, in fact, satisfy the homogeneous equation.

***Homogeneous Solution*** $f_n = 0$, ***Repeated Roots:*** $r_1 = r_2 = r$     The homogeneous solution for this case is the sum of two exponential functions $C_1 r^n$ and $C_2 n r^n$. Substituting $C_1 r^n$ into (5.24) gives

$$(D-r)(D-r)C_1 r^n \overset{?}{=} 0$$

Using (5.25a) shows immediately that

$$(D-r)\underbrace{(D-r)C_1 r^n}_{0} \overset{?}{=} 0$$

Similarly, substituting $C_2 n r^n$ into (5.24) and using (5.25b) gives

$$(D-r)\underbrace{(D-r)C_2 n r^n}_{C_2 r^{(n+1)}} \overset{?}{=} 0$$

Operating on this with the second operator gives, using (5.25a),

$$\underbrace{(D-r)C_2 r^{(n+1)}}_{0} = 0$$

Again the difference operator has provided an easy and immediate means of demonstrating that these assumed forms, in fact, satisfy the homogeneous equation.

***The Forced Solution*** $f_n \neq 0$ ***for Special Cases***     Determining the forced solution is particularly simple using the *method of undetermined coefficients*: Assume a *form* of the forced solution that is the same *form* as $f_n$. But there were two special cases where this method required modification. The first case was where the roots of the characteristic equation are distinct and $f_n = K r_i^n$, with $r_i$ being a root of the characteristic equation. If we assume the usual form of the forced solution as $y_n^{(f)} = M r_i^n$, this could not be made to satisfy the nonhomogeneous equation because that form was the same form as in the homogeneous solution. We can easily demonstrate this with the difference operator:

$$(D-r_2)\underbrace{(D-r_1)M r_1^n}_{0} \neq K r_1^n$$

$$(D-r_1)\underbrace{(D-r_2)M r_2^n}_{0} \neq K r_2^n$$

where we used the result in (5.25a). For this case we multiplied the usual assumed form of the forced solution by $n$ to give $y_n^{(f)} = M n r_i^n$. Using the difference operator, we can easily show that this form can be made to satisfy the nonhomogeneous equation

$$(D - r_2)\underbrace{(D - r_1)Mnr_1^n}_{Mr_1^{(n+1)}} = r_1(r_1 - r_2)Mr_1^n = Kr_1^n$$

$$(D - r_1)\underbrace{(D - r_2)Mnr_2^n}_{Mr_2^{(n+1)}} = r_2(r_2 - r_1)Mr_2^n = Kr_2^n$$

where we used the result in (5.25b), thereby giving a solution for $M$ in either case.

If the roots of the characteristic equation are repeated, $r_1 = r_2 = r$, and $f_n = Kr^n$, assuming a form of the forced solution of $y_n^{(f)} = Mnr^n$ would not satisfy the nonhomogeneous equation since $r_1 = r_2$ and the left-hand sides of the previous result would be zero. This is again rather obvious because $y_n^{(f)} = Mnr^n$ has the same form as a part of the homogeneous solution for this case of repeated roots and satisfies the homogeneous equation whose right-hand side is zero. So we again multiplied this by $n$ and assumed the form of the forced solution as $y_n^{(f)} = Mn^2r^n$. Using the difference operator, we can easily show that this form can be made to satisfy the nonhomogeneous equation

$$(D - r)\underbrace{(D - r)Mn^2r^n}_{(2n+1)Mr^{(n+1)}} = 2r^2 Mr^n = Kr^n$$

where we used the basic result in (5.25c).

# 6 Solution of Linear, Constant-Coefficient, Partial Differential Equations

In this chapter we study the solution of *partial differential equations* (PDEs), which contain partial derivatives. In previous chapters the variables were functions of only one independent variable [e.g., $x(t)$, $y(x)$]. Hence, only *ordinary derivatives* of those variables were required [e.g., $dx(t)/dt$, $dy(x)/dx$]. Partial differential equations involve variables that are functions of *more than one independent variable*: for example, $u(x, t)$ and $v(y, x)$. Hence, we need *partial derivatives* such as $\partial u(x, t)/\partial x$, $\partial u(x, t)/\partial t$, $\partial v(y, x)/\partial y$, and $\partial v(y, x)/\partial x$. The key idea in the solution of PDEs is a very common mathematical principle: We try to reduce them to differential equations that we already know how to solve, such as ordinary differential equations or difference equations.

## 6.1 COMMON ENGINEERING PARTIAL DIFFERENTIAL EQUATIONS

There are several PDEs that commonly appear throughout all disciplines of engineering.

### Laplace's and Poisson's Equation

$$\frac{\partial^2 \phi(x, y)}{\partial x^2} + \frac{\partial^2 \phi(x, y)}{\partial y^2} = \rho(x, y)$$

This is one of the most common PDEs in electromagnetics in electrical engineering. It gives the voltage $\phi(x, y)$ in a two-dimensional space, $x$, $y$, that is caused by a charge distribution over that space, $\rho(x, y)$. It is referred to as Laplace's equation if $\rho(x, y) = 0$.

---

*Essential Math Skills for Engineers*, By Clayton R. Paul
Copyright © 2009 John Wiley & Sons, Inc.

## The Wave Equation

$$\boxed{\frac{\partial^2 \phi(x,t)}{\partial x^2} = \frac{1}{v^2} \frac{\partial^2 \phi(x,t)}{\partial t^2}}$$

This is also one of the most common PDEs in electromagnetics in electrical engineering. It describes the propagation of voltage, electric field, magnetic field, and so on, *waves* $\phi(x, t)$ in terms of their position $x$ and time $t$, where $v$ is the velocity of propagation of the wave. It is also very common in mechanical engineering, giving, for example, the displacement of a vibrating string versus time, the torsional vibration of a shaft, and so on. For two dimensions,

$$\boxed{\frac{\partial^2 \phi(x,y,t)}{\partial x^2} + \frac{\partial^2 \phi(x,y,t)}{\partial y^2} = \frac{1}{v^2} \frac{\partial^2 \phi(x,y,t)}{\partial t^2}}$$

gives the position in two dimensions, $x$ and $y$, of a vibrating membrane versus time.

## The Vibrating Beam Equation

$$\boxed{\frac{\partial^2 y(x,t)}{\partial t^2} + \frac{EI}{A\rho} \frac{\partial^4 y(x,t)}{\partial x^4} = 0}$$

This is common in mechanical engineering, giving the transverse vibrations of a beam versus time. $E$ is Young's modulus of elasticity, $I$ is the cross-sectional moment of inertia, $A$ is the beam cross-sectional area, $\rho$ is the density, and $y$ is the beam deflection.

## The Diffusion Equation or Heat Flow Equation

$$\boxed{\frac{\partial^2 \phi(x,t)}{\partial x^2} = k^2 \frac{\partial \phi(x,t)}{\partial t}}$$

This is common in electrical engineering, showing how the current in a good conductor diffuses into that conductor over time. It is also common in mechanical engineering, giving the heat flow in a material that has a thermal conductivity.

## 6.2  THE LINEAR, CONSTANT-COEFFICIENT, PARTIAL DIFFERENTIAL EQUATION

The most general form of a *second-order, linear, constant-coefficient, partial differential equation* involves all of the derivatives, up through the second derivative, with respect to two independent variables, denoted here as $x$ and $y$:

$$A\frac{\partial^2 u(x, y)}{\partial x^2} + B\frac{\partial^2 u(x, y)}{\partial x \partial y} + C\frac{\partial^2 u(x, y)}{\partial y^2}$$
$$+ D\frac{\partial u(x, y)}{\partial x} + E\frac{\partial u(x, y)}{\partial y} + Fu(x, y) = f(x, y)$$

(6.1)

The equation is said to be *linear* since neither the variable $u(x, y)$ nor its derivatives are raised to a power greater than unity and are not multiplied times each other. The equation is *constant-coefficient* since none of the coefficients $A$, $B$, $C$, $D$, $E$, or $F$ are functions of the independent variables $x$ and $y$. Linear and constant-coefficient PDEs represent the most common and useful PDEs in all disciplines of engineering.

## 6.3   THE METHOD OF SEPARATION OF VARIABLES

The *method of separation of variables* is the most general method for solving PDEs and works for virtually all PDEs of engineering interest. Suppose that the variable to be solved for, $u$, is a function of two independent variables, $x$ and $y$, as $u(x, y)$. We assume a form of the solution that is the *product of two functions*, each of which is a function of only one of the independent variables as

$$u(x, y) = X(x)Y(y)$$

(6.2)

Here $X(x)$ denotes a function of $x$ only and is independent of $y$. Similarly, $Y(y)$ denotes a function of $y$ only and is independent of $x$. Substituting (6.2) into the PDE will give *two separate ordinary differential equations*, which we already know how to solve. This represents an important technique that is commonly used in mathematics: We try to reduce a new problem to a form that we already know how to solve.

For example, consider Laplace's equation:

$$\frac{\partial^2 \phi(x, y)}{\partial x^2} + \frac{\partial^2 \phi(x, y)}{\partial y^2} = 0$$

(6.3)

We assume that $\phi(x, y)$ can be written as a product of $X(x)$ which is independent of $y$ and $Y(y)$ which is independent of $x$ as

$$\phi(x, y) = X(x)Y(y)$$

(6.4)

giving

$$\frac{\partial^2 X(x)}{\partial x^2}Y(y) + \frac{\partial^2 Y(y)}{\partial y^2}X(x) = 0$$

Here is the *heart of the method of separation of variables*: Divide both sides by the product $X(x)Y(y)$, to give

$$\frac{1}{X(x)}\frac{d^2 X(x)}{dx^2} + \frac{1}{Y(y)}\frac{d^2 Y(y)}{dy^2} = 0$$

The first term is *independent of y*, and the second term is *independent of x*. The only way that their sum can equal zero is if *each term is a constant independent of x and y!* So we write

$$\frac{1}{X(x)}\frac{d^2 X(x)}{dx^2} = k_x^2$$

$$\frac{1}{Y(y)}\frac{d^2 Y(y)}{dy^2} = -k_y^2$$

We have squared these constants knowing ahead of time that this will simplify the later solution. Since the right-hand side of the PDE is zero, the sum of these constants must also be zero, so we have chosen one of the constants to be negative. Therefore, the two constants are related:

$$k_x^2 - k_y^2 = 0$$

Hence, the magnitudes of both constants must be equal, so we really have only one undetermined constant $k$:

$$k_x^2 = k^2$$
$$k_y^2 = k^2$$

So we have determined how to write the original PDE as two *ordinary differential equations*:

$$\boxed{\frac{d^2 X(x)}{dx^2} - k^2 X(x) = 0} \tag{6.5a}$$

$$\boxed{\frac{d^2 Y(y)}{dy^2} + k^2 Y(y) = 0} \tag{6.5b}$$

As was determined in Chapter 4, the solution to each of these ordinary differential equations is just the homogeneous solution. The roots of the characteristic equation for $X(x)$ in (6.5a) are $p_1, p_2 = \pm k$. Hence, its general solution is

$$X(x) = X_1 e^{kx} + X_2 e^{-kx}$$

where $X_1$ and $X_2$ are two (as yet) undetermined constants. The roots of the characteristic equation for $Y(y)$ in (6.5b) are $p_1, p_2 = \pm jk$. Hence, its general solution is

$$Y(y) = Y_1 \cos ky + Y_2 \sin ky$$

where $Y_1$ and $Y_2$ are two (as yet) undetermined constants. (You should "sanity check" these results.) The total solution to the original PDE is the product of these solutions:

$$\boxed{\begin{aligned} \phi(x, y) &= X(x) Y(y) \\ &= (X_1 e^{kx} + X_2 e^{-kx})(Y_1 \cos ky + Y_2 \sin ky) \end{aligned}} \tag{6.6}$$

The foregoing is the heart of the *method of separation of variables*. The *form* of the general solution to virtually all PDEs of engineering interest can be obtained with this simple technique. It now remains to determine these four undetermined constants, $X_1$, $X_2$, $Y_1$, and $Y_2$, as well as the constant $k$ in that general solution. This will be done using the *boundary conditions* for the specific problem.

## 6.4   BOUNDARY CONDITIONS AND INITIAL CONDITIONS

Using the method of separation of variables is the easy part of the solution process. The main difficulty arises in the application of the *boundary conditions* for the specific problem to determine the four undetermined constants, $X_1$, $X_2$, $Y_1$, and $Y_2$, as well as the constant $k$ in the general solution. But this remaining part of the solution process is rather simple in concept as long as you are willing to think and be clever.

As the name implies, the *boundary conditions* are the specifications, *for the specific problem*, of the values of the variable, for example, $u(x, y)$, at the physical *boundaries* of the problem. When the independent variables are denoted as $x$ or $y$, this implies that they are dimensional variables such as position coordinates of a rectangular coordinate system—hence the name *boundary conditions*. We might specify the value of $u(x, y)$ at a point $x = 0$ and $y = 2$ by specifying

$$u(x = 0, y = 2) \equiv u(0, 2)$$

For a second-order PDE, we need *four* such boundary conditions to determine the four undetermined constants in the general solution: $X_1$, $X_2$, $Y_1$, and $Y_2$. For some problems one of the independent variables may be time denoted as $t$, such as $u(x, t)$. Specifying $u(x, t)$ at the start and end values of time for a

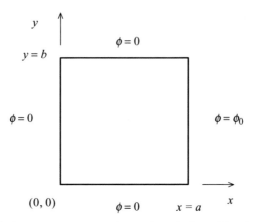

**Fig. 6.1.** Specifying the boundary voltages for solution of Laplace's equation.

problem makes them equivalent to "boundary conditions," but they are called *initial conditions*.

Now here is where we have to start being clever. For example, consider the second-order Laplace's equation whose general solution we obtained previously:

$$\boxed{\frac{\partial^2 \phi(x,y)}{\partial x^2} + \frac{\partial^2 \phi(x,y)}{\partial y^2} = 0} \tag{6.3}$$

Suppose that the actual problem consists of four metal plates on which the voltage, $\phi(x, y)$, is specified as shown in Figure 6.1. The voltage of the right side is specified as $\phi = \phi_0$, and the voltages of the other three sides are specified as zero. Hence, the boundary conditions for this specific problem are specified as

$$\phi(x, 0) = 0$$
$$\phi(x, b) = 0$$
$$\phi(0, y) = 0$$
$$\phi(a, y) = \phi_0 \tag{6.7}$$

The general solution to Laplace's equation was determined by the method of separation of variables as

$$\phi(x,y) = (X_1 e^{kx} + X_2 e^{-kx})(Y_1 \cos ky + Y_2 \sin ky) \tag{6.6}$$

Applying the first boundary condition gives

$$\phi(x,0) = (X_1 e^{kx} + X_2 e^{-kx}) \left[ Y_1 \underbrace{\cos(k \cdot 0)}_{1} + Y_2 \underbrace{\sin(k \cdot 0)}_{0} \right]$$
$$= Y_1(X_1 e^{kx} + X_2 e^{-kx})$$
$$= 0$$

The only way that this can be satisfied for all $x$ is for $Y_1 = 0$, leaving

$$\phi(x, y) = (X_1 e^{kx} + X_2 e^{-kx})(Y_2 \sin ky)$$

Applying the next boundary condition gives

$$\phi(x,b) = (X_1 e^{kx} + X_2 e^{-kx})(Y_2 \sin kb)$$
$$= 0$$

We can't make $Y_2 = 0$ because that would render the entire solution zero. But we can choose

$$kb = n\pi \qquad \text{for} \quad n = 0, 1, 2, \ldots$$

since $\sin n\pi = 0$ for $n = 0, 1, 2, \ldots$ . Therefore,

$$k = \frac{n\pi}{b} \tag{6.8}$$

and we have determined the constant $k$. Next, we apply the third boundary condition:

$$\phi(0, y) = (X_1 + X_2)\left(Y_2 \sin \frac{n\pi y}{b}\right)$$
$$= 0$$

This means that $X_2 = -X_1$ and we have

$$\phi(x, y) = (X_1 e^{n\pi x/b} - X_1 e^{-n\pi x/b})\left[ Y_2 \sin \frac{n\pi y}{b} \right]$$
$$= X_1 Y_2 [e^{n\pi x/b} - e^{-n\pi x/b}] \sin \frac{n\pi y}{b}$$
$$= C \sinh \frac{n\pi x}{b} \sin \frac{n\pi y}{b} \tag{6.9}$$

and we have combined the two undetermined constants into one: $C = 2X_1 Y_2$. We have also substituted the relation for the hyperbolic sine from Chapter 2:

$$\sinh \theta = \frac{e^{\theta} - e^{-\theta}}{2}$$

We now have one remaining boundary condition to impose, $\phi(a,y) = \phi_0$. Applying this to (6.9) yields

$$\phi(a,y) = \phi_0 = C \sinh \frac{n\pi a}{b} \sin \frac{n\pi y}{b}$$

But this cannot be satisfied *for all* y with only one unknown C. However, we can add an infinite number of such solutions to give

$$\boxed{\phi(x,y) = \sum_{n=1}^{\infty} C_n \sinh \frac{n\pi x}{b} \sin \frac{n\pi y}{b}} \qquad (6.10)$$

Now applying the fourth boundary condition gives

$$\phi(a,y) = \sum_{n=1}^{\infty} C_n \sinh \frac{n\pi a}{b} \sin \frac{n\pi y}{b}$$
$$= \phi_0$$

This is in the form of a Fourier sine series (see Chapter 7):

$$\phi_0 = \sum_{n=1}^{\infty} \underbrace{C_n \sinh \frac{n\pi a}{b}}_{a_n} \sin \frac{n\pi y}{b}$$

where the coefficients of the Fourier sine series, $a_n$, are determined from the Fourier series methods of Chapter 7 as

$$a_n = \frac{2}{b} \int_0^b \phi_0 \sin \frac{n\pi y}{b} \, dy$$
$$= -\phi_0 \frac{2}{b}\left(\frac{b}{n\pi}\right) \cos \frac{n\pi y}{b}\Big|_0^b$$
$$= \begin{cases} 0 & n \text{ even} \\ \dfrac{4\phi_0}{n\pi} & n \text{ odd} \end{cases}$$

Therefore, $C_n$ is

$$C_n = \frac{a_n}{\sinh n\pi a/b}$$
$$= \frac{4\phi_0}{n\pi \sinh n\pi a/b} \qquad n \text{ odd} \qquad (6.11)$$

This gives the final, complete solution as

$$\phi(x,y) = \sum_{\substack{n=1 \\ n \text{ odd}}}^{\infty} \frac{4\phi_0}{n\pi} \frac{\sinh n\pi x/b}{\sinh n\pi a/b} \sin \frac{n\pi y}{b} \qquad (6.12)$$

Granted, this final step, imposition of the boundary conditions, is a bit tedious, but that is generally what happens in the solution of PDEs.

---

### Example

Solve the wave equation

$$\frac{\partial^2 u(x,t)}{\partial x^2} = \frac{1}{v^2}\frac{\partial^2 u(x,t)}{\partial t^2} \qquad (6.13)$$

Solve this by assuming, by separation of variables, that

$$u(x,t) = X(x)T(t) \qquad (6.14)$$

Substituting yields

$$\frac{d^2 X(x)}{dx^2}T(t) = \frac{1}{v^2}\frac{d^2 T(t)}{dt^2}X(x)$$

Dividing both sides by the product $X(x)T(t)$ gives

$$\frac{1}{X(x)}\frac{d^2 X(x)}{dx^2} = \frac{1}{v^2}\frac{1}{T(t)}\frac{d^2 T(t)}{dt^2}$$

The left side is independent of $t$ and the right side is independent of $x$, so both sides must equal the same constant (which we arbitrarily square and choose as negative):

$$\frac{1}{X(x)}\frac{d^2 X(x)}{dx^2} = -k^2$$

$$\frac{1}{v^2}\frac{1}{T(t)}\frac{d^2 T(t)}{dt^2} = -k^2$$

Note that this result is different from that of Laplace's equation solved previously. In that problem the left side was the sum of the two functions and was equal to zero. Hence, in that solution the constants were the negative of each other.

Again we obtain two ordinary differential equations to solve:

$$\boxed{\frac{d^2X(x)}{dx^2} + k^2X(x) = 0}$$  (6.15a)

$$\boxed{\frac{d^2T(t)}{dt^2} + v^2k^2T(t) = 0}$$  (6.15b)

The general solutions are

$$X(x) = X_1 \cos kx + X_2 \sin kx$$

$$T(t) = T_1 \cos vkt + T_2 \sin vkt$$

giving the total solution as

$$\boxed{u(x,t) = (X_1 \cos kx + X_2 \sin kx)(T_1 \cos vkt + T_2 \sin vkt)}$$  (6.16)

Up to this point the solution process is the same as before.

Now we apply the boundary and initial conditions. To formulate these we must consider a specific problem. We use the example of a vibrating string shown in Figure 6.2.

The variable $u(x, t)$ gives the vertical position of points on the string as it vibrates up and down. The string is fixed at the two endpoints, $x = 0$ and $x = L$, giving the first two boundary conditions:

$$u(0, t) = 0$$

$$u(L, t) = 0$$

Applying these to the general solution gives

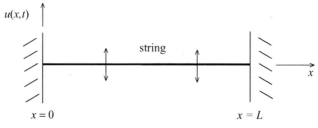

**Fig. 6.2.** A vibrating string.

$$u(0, t) = \left[ X_1 \underbrace{\cos(k \cdot 0)}_{1} + X_2 \underbrace{\sin(k \cdot 0)}_{0} \right] (T_1 \cos vkt + T_2 \sin vkt)$$
$$= 0$$

which shows that $X_1 = 0$ and

$$u(L, t) = X_2 \sin kL (T_1 \cos vkt + T_2 \sin vkt)$$
$$= 0$$

which shows that

$$kL = n\pi \qquad \text{for} \quad n = 0, 1, 2, 3, \ldots$$

or

$$k = \frac{n\pi}{L} \tag{6.17}$$

Thus, the solution so far is

$$u(x, t) = X_2 \sin \frac{n\pi x}{L} \left[ T_1 \cos\left(v \frac{n\pi t}{L}\right) + T_2 \sin\left(v \frac{n\pi t}{L}\right) \right]$$

The remaining conditions are the "initial conditions." We could specify the initial displacement of the points along the string (as a function of $x$) at $t = 0$ as

$$u(x, 0) = U_0(x)$$

and the initial velocity of points along the string (as a function of $x$) at $t = 0$ as

$$\left. \frac{du}{dt} \right|_{t=0} = V_0(x)$$

Again we add an infinite number of solutions to give

$$u(x, t) = \sum_{n=1}^{\infty} \sin \frac{n\pi x}{L} \left[ C_{n1} \cos\left(v \frac{n\pi t}{L}\right) + C_{n2} \sin\left(v \frac{n\pi t}{L}\right) \right] \tag{6.18}$$

Evaluating with these two initial conditions gives

$$u(x, 0) = U_0(x)$$

$$= \sum_{n=1}^{\infty} \sin\frac{n\pi x}{L}\left(C_{n1}\underbrace{\cos 0}_{1} + C_{n2}\underbrace{\sin 0}_{0}\right)$$

$$= \sum_{n=1}^{\infty} C_{n1} \sin\frac{n\pi x}{L}$$

Again this is a Fourier sine series with the coefficient given as

$$C_{n1} = \frac{2}{L}\int_0^L U_0(x)\sin\frac{n\pi x}{L}\,dx$$

Given the initial string position as a function of $x$, we can evaluate this integral to yield $C_{n1}$.

Applying the last initial condition gives

$$\left.\frac{du(x, t)}{dt}\right|_{t=0} = \sum_{n=1}^{\infty}\left(v\frac{n\pi}{L}\right)\sin\frac{n\pi x}{L}\left(-C_{n1}\underbrace{\sin 0}_{0} + C_{n2}\underbrace{\cos 0}_{1}\right)$$

$$= \sum_{n=1}^{\infty} C_{n2}\left(v\frac{n\pi}{L}\right)\sin\frac{n\pi x}{L}$$

$$= V_0(x)$$

Again this is a Fourier sine series, so that $C_{n2}$ is determined by

$$C_{n2} = \frac{L}{v n\pi}\left(\frac{2}{L}\right)\int_0^L V_0(x)\sin\frac{n\pi x}{L}\,dx$$

Given the initial string velocity as a function of $x$, we can evaluate this integral to yield $C_{n2}$ and the solution is complete:

$$\boxed{u(x, t) = \sum_{n=1}^{\infty}\sin\frac{n\pi x}{L}\left[C_{n1}\cos\left(v\frac{n\pi t}{L}\right) + C_{n2}\sin\left(v\frac{n\pi t}{L}\right)\right]}$$

Again the only difficult part of the solution process is the incorporation of the boundary and initial conditions.

This solution can be rewritten in a more revealing form using the identities obtained in Chapter 2:

$$\sin A \cos B = \frac{1}{2}\sin(A + B) + \frac{1}{2}\sin(A - B)$$

$$\sin A \sin B = \frac{1}{2}\cos(A - B) - \frac{1}{2}\cos(A + B)$$

giving an equivalent form of the solution:

$$u(x,t) = \frac{1}{2}\sum_{n=1}^{\infty} C_{n1}\left[\sin\left(\frac{n\pi}{L}(x+vt)\right) + \sin\left(\frac{n\pi}{L}(x-vt)\right)\right]$$
$$+ \frac{1}{2}\sum_{n=1}^{\infty} C_{n2}\left[\cos\left(\frac{n\pi}{L}(x-vt)\right) - \cos\left(\frac{n\pi}{L}(x+vt)\right)\right]$$

Notice that the solutions consist of functions that involve position $x$, wave velocity $v$, and time $t$ only as $x - vt$ and $x + vt$. These are traveling waves and the general form of the solution can be written as

$$u(x,t) = \sum_{n=1}^{\infty} K_n^+ f_n(x-vt) + \sum_{n=1}^{\infty} K_n^- f_n(x+vt) \qquad (6.19)$$

The functions $f_n(x - vt)$ are said to be forward-traveling waves (i.e., traveling in the $+x$ direction) since as $t$ increases, $x$ must also increase to keep the argument of the function constant (i.e., to track a point on the wave). For similar reasons, the functions $f_n(x + vt)$ are said to be backward-traveling waves (i.e., traveling in the $-x$ direction). (Verify, by direct substitution, that this general form satisfies the wave equation.)

---

## 6.5 NUMERICAL (COMPUTER) SOLUTIONS VIA FINITE DIFFERENCES: CONVERSION TO DIFFERENCE EQUATIONS

The analytical solution of partial differential equations is a bit tedious when we apply the boundary conditions. However, they can be solved much more easily in an approximate fashion via *finite differences*. This amounts to the same technique as that used in Chapter 4 to solve ordinary differential equations. That is, we discretize the variables and solve only at discrete points via a difference equation.

For example, consider Laplace's equation:

$$\frac{\partial^2 \phi(x,y)}{\partial x^2} + \frac{\partial^2 \phi(x,y)}{\partial y^2} = 0$$

We can approximate partial derivatives in the following manner. The first partial derivative at a point midway between points $P$ and $B$ shown in Figure 6.3 with respect to $x$ is approximated as

$$\left.\frac{\partial f(x,y)}{\partial x}\right|_{x=x_0+\Delta x/2} \cong \frac{f(x_0 + \Delta x, y) - f(x_0, y)}{\Delta x}$$

giving the slope of the line $PB$. Similarly, the first partial derivative at a point midway between points $A$ and $P$ with respect to $x$ is approximated as

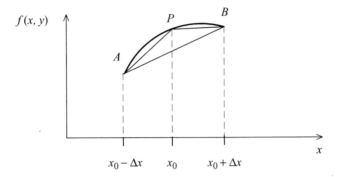

**Fig. 6.3.** Approximating partial derivatives.

$$\frac{\partial f(x, y)}{\partial x}\bigg|_{x=x_0-\Delta x/2} \cong \frac{f(x_0, y)-f(x_0-\Delta x, y)}{\Delta x}$$

giving the slope of the line $AP$. The second partial derivative at point $P$ can then be approximated as

$$\frac{\partial^2 f}{\partial x^2}\bigg|_{x=x_0} \cong \frac{1}{\Delta x}\left(\frac{\partial f}{\partial x}\bigg|_{x=x_0+\Delta x/2} - \frac{\partial f}{\partial x}\bigg|_{x=x_0-\Delta x/2}\right)$$
$$= \frac{f(x_0+\Delta x, y)-2f(x_0, y)+f(x_0-\Delta x, y)}{\Delta x^2}$$

(6.20)

This is called a central-difference approximation.

To approximate Laplace's equation, we discretize the $x$ and $y$ space variables as $\Delta x$ and $\Delta y$. Denote the voltage at these points as

$$\phi_{i,j} \equiv \phi(i\,\Delta x, j\,\Delta y)$$

(6.21)

The second partial derivatives, using the results above, are denoted as

$$\frac{\partial^2 \phi}{\partial x^2}\bigg|_{i\Delta x, j\Delta y} \cong \frac{\phi_{i+1,j}-2\phi_{i,j}+\phi_{i-1,j}}{\Delta x^2}$$

(6.22a)

and

$$\frac{\partial^2 \phi}{\partial y^2}\bigg|_{i\Delta x, j\Delta y} \cong \frac{\phi_{i,j+1}-2\phi_{i,j}+\phi_{i,j-1}}{\Delta y^2}$$

(6.22b)

Hence Laplace's equation is approximated as

$$\boxed{\frac{\phi_{i+1,j} - 2\phi_{i,j} + \phi_{i-1,j}}{\Delta x^2} + \frac{\phi_{i,j+1} - 2\phi_{i,j} + \phi_{i,j-1}}{\Delta y^2} = 0} \qquad (6.23)$$

To simplify the result we will assume that the $x$ and $y$ discretizations are equal:

$$\Delta x = \Delta y = h$$

Laplace's equation becomes

$$\boxed{\phi_{i,j} = \frac{1}{4}(\phi_{i+1,j} + \phi_{i-1,j} + \phi_{i,j+1} + \phi_{i,j-1}) \qquad \Delta x = \Delta y = h} \qquad (6.24)$$

This remarkable result shows that *the voltage at a point is the average of the voltages at the four adjacent points*, as indicated in Figure 6.4!

---

### *Example*

The voltages at the four sides of a rectangular region are fixed (known) as shown in Figure 6.5, and we want to solve for the resulting voltages at six interior points $a$, $b$, $c$, $d$, $e$, and $f$ so that these satisfy Laplace's equation.

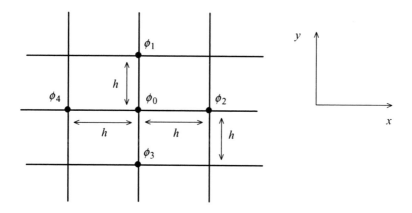

$$\phi_0 = \frac{1}{4}(\phi_1 + \phi_2 + \phi_3 + \phi_4) \qquad \Delta x = \Delta y = h$$

**Fig. 6.4.** Solution of Laplace's equation by finite differences.

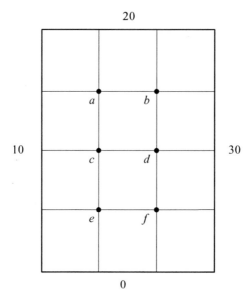

**Fig. 6.5.** Solution of Laplace's equation by iteration.

One way to solve this is by *iteration*. The voltages at the interior points are related by (6.24)

$$\phi_a = \frac{1}{4}(20 + \phi_b + \phi_c + 10)$$

$$\phi_b = \frac{1}{4}(20 + 30 + \phi_d + \phi_a)$$

$$\phi_c = \frac{1}{4}(\phi_a + \phi_d + \phi_e + 10)$$

$$\phi_d = \frac{1}{4}(\phi_b + 30 + \phi_f + \phi_c)$$

$$\phi_e = \frac{1}{4}(\phi_c + \phi_f + 0 + 10)$$

$$\phi_f = \frac{1}{4}(\phi_d + 30 + 0 + \phi_e)$$

First assume that the voltages at the interior points are zero:

$$\phi_a = \phi_b = \phi_c = \phi_d = \phi_e = \phi_f = 0$$

Then compute the voltages at the nodes in succession, $a \to b \to c \to d \to e \to f$, using the voltages that are computed for prior nodes in that cycle. For example, for the first cycle of computation,

$$\phi_a = \frac{1}{4}(20+0+0+10) = 7.5$$

$$\phi_b = \frac{1}{4}(20+30+0+7.5) = 14.38$$

$$\phi_c = \frac{1}{4}(7.5+0+0+10) = 4.38$$

$$\phi_d = \frac{1}{4}(14.38+30+0+4.38) = 12.19$$

$$\phi_e = \frac{1}{4}(4.38+0+0+10) = 3.59$$

$$\phi_f = \frac{1}{4}(12.19+30+0+3.59) = 11.45$$

After 10 cycles we obtain $\phi_a = 16.44$, $\phi_b = 21.66$, $\phi_c = 14.10$, $\phi_d = 20.19$. $\phi_e = 9.77$, and $\phi_f = 14.99$.

The formulas above for the potentials of the interior nodes can also be placed in matrix form as

$$\underbrace{\begin{bmatrix} -4 & 1 & 1 & 0 & 0 & 0 \\ 1 & -4 & 0 & 1 & 0 & 0 \\ 1 & 0 & -4 & 1 & 1 & 0 \\ 0 & 1 & 1 & -4 & 0 & 1 \\ 0 & 0 & 1 & 0 & -4 & 1 \\ 0 & 0 & 0 & 1 & 1 & -4 \end{bmatrix}}_{\mathbf{A}} \underbrace{\begin{bmatrix} \phi_a \\ \phi_b \\ \phi_c \\ \phi_d \\ \phi_e \\ \phi_f \end{bmatrix}}_{\mathbf{X}} = - \underbrace{\begin{bmatrix} 30 \\ 50 \\ 10 \\ 30 \\ 10 \\ 30 \end{bmatrix}}_{\mathbf{B}}$$

This matrix equation can then be solved directly by the methods of Chapter 3:

$$\mathbf{X} = \mathbf{A}^{-1}\mathbf{B}$$

The following table compares the results of the iteration method (for 10 iterations), the direct solution of the matrix equation, and the exact solution (by the analytical solution of Laplace's equation using the method of separation of variables):

|          | Exact  | Matrix | Iteration |
|----------|--------|--------|-----------|
| $\phi_a$ | 16.48  | 16.44  | 16.44     |
| $\phi_b$ | 21.85  | 21.66  | 21.66     |
| $\phi_c$ | 14.16  | 14.10  | 14.10     |
| $\phi_d$ | 20.49  | 20.19  | 20.19     |
| $\phi_e$ | 9.61   | 9.77   | 9.77      |
| $\phi_f$ | 14.98  | 14.99  | 14.99     |

***Example***

Finally, we solve the diffusion equation or heat flow equation:

$$\frac{\partial \phi(x,t)}{\partial t} = \frac{1}{k^2} \frac{\partial^2 \phi(x,t)}{\partial x^2} \tag{6.25}$$

using this method. Substituting the difference approximations above for the partial derivatives gives

$$\frac{\phi_{i,j+1} - \phi_{i,j}}{\Delta t} = \frac{1}{k^2} \frac{\phi_{i+1,j} - 2\phi_{i,j} + \phi_{i-1,j}}{\Delta x^2} \tag{6.26a}$$

where

$$\phi_{i,j} \equiv \phi(i\,\Delta x, j\,\Delta t) \tag{6.26b}$$

giving

$$\phi_{i,j+1} = \alpha\phi_{i+1,j} + (1 - 2\alpha)\phi_{i,j} + \alpha\phi_{i-1,j} \tag{6.27a}$$

where

$$\alpha = \frac{\Delta t}{k^2 \Delta x^2} \tag{6.27b}$$

Unlike Laplace's equation, which is stable regardless of the size of the discretizations, $\Delta x$ and $\Delta y$, this solution is stable only for $\Delta x$ and $\Delta t$ satisfying

$$0 < \alpha = \frac{\Delta t}{k^2 \Delta x^2} \leq \frac{1}{2}$$

Assume the boundary conditions to be

$$\phi(0,t) = 0 = \phi(L,t) = 0 \qquad \text{for all } t > 0$$

and the initial condition prescribes the temperature at $t = 0$ as a function of position along the bar:

$$\phi(x, 0) = \Phi(x)$$

This is equivalent to a one-dimensional heat flow through a bar of length $L$ with both ends held at a temperature of zero degrees. Choose $\Delta x$ and $\Delta t$ by dividing the bar length into $n$ equal-length segments:

$$\Delta x = \frac{L}{n}$$

and dividing the total solution time, $T_{final}$, into $m$ equal time increments:

$$\Delta t = \frac{T_{final}}{m}$$

such that the stability condition above is satisfied:

$$\Delta t \leq \frac{1}{2}k^2 \Delta x^2$$

or

$$n^2 \leq \left(\frac{1}{2}k^2 \frac{L^2}{T_{final}}\right) m \tag{6.28}$$

Solution of the difference equation in (6.27) gives the temperature of the bar as a function of time and position along its length. Start solving first at $t = 0$ $(j = 0)$ and incrementing position along the bar, $i = 0, 1, 2, \dots, n$:

$$\phi_{i,1} = \alpha\phi_{i+1,0} + (1 - 2\alpha)\phi_{i,0} + \alpha\phi_{i-1,0}$$

Then increment for $t = \Delta t$ $(j = 1)$ and re-solve by incrementing position along the bar again:

$$\phi_{i,2} = \alpha\phi_{i+1,1} + (1 - 2\alpha)\phi_{i,1} + \alpha\phi_{i-1,1}$$

and so on.

# 7 The Fourier Series and Fourier Transform

All engineering systems can be represented, *symbolically*, by a *block diagram*, as shown in Figure 7.1. One or more *inputs* represented by $f(t)$ are applied to the system. The system *processes* the inputs producing one or more *outputs* represented by $y(t)$. The inputs and outputs are shown as being a function of $t$, which usually denotes time.

The relationship between the input and the output is generally described (governed) by some form of mathematical equation, such as an algebraic equation (Chapter 3), an ordinary differential equation (Chapter 4), a difference equation (Chapter 5), or a partial differential equation (Chapter 6). For example, consider the electric circuit shown in Figure 1.2. The input is the voltage source, $V_S(t)$, and the output is the voltage across the capacitor, $V(t)$. The input and output are related by the second-order, linear, constant-coefficient, ordinary differential equation

$$\frac{d^2V(t)}{dt^2} + \frac{R}{L}\frac{dV(t)}{dt} + \frac{1}{LC}V(t) = \frac{1}{LC}V_S(t)$$

We have seen that solving some of those governing equations can be somewhat tedious. The heart of this chapter is the concept of *decomposing* the input into a sum of more basic waveforms, $f_i(t)$, as

$$\begin{aligned} f(t) &= f_0 + \sum_{i=1}^{n} f_i(t) \\ &= f_0 + f_1(t) + f_2(t) + \cdots + f_n(t) \end{aligned} \tag{7.1}$$

If the engineering system is *linear*, the output is the sum of the responses to the individual inputs as

$$\begin{aligned} y(t) &= y_0 + \sum_{i=1}^{n} y_i(t) \\ &= y_0 + y_1(t) + y_2(t) + \cdots + y_n(t) \end{aligned} \tag{7.2}$$

*Essential Math Skills for Engineers*, By Clayton R. Paul
Copyright © 2009 John Wiley & Sons, Inc.

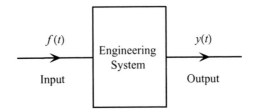

**Fig. 7.1.** All engineering systems have an input and an output.

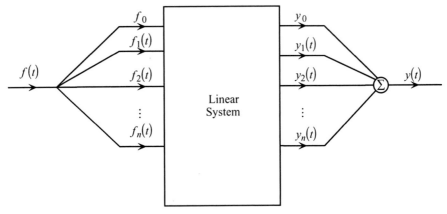

**Fig. 7.2.** Using superposition for a linear system to determine the output for a complicated waveform.

where $f_0$ and $y_0$ are constants and each component of the input, $f_i(t)$, produces a corresponding component of the output, $y_i(t)$, as

$$\boxed{f_i(t) \rightarrow y_i(t) \qquad i = 0, 1, 2, \ldots, n}\tag{7.3}$$

which is illustrated in Figure 7.2.

This is known as the principle of *superposition* and is fundamental to all *linear* systems. Linear systems are described by linear equations (algebraic or differential). If we carefully choose the $f_i(t)$ waveforms to be some basic ones for which we can more easily determine the output, the original objective of determining the output $y(t)$ in response to the original input $f(t)$ can be reduced to determining the responses to the individual basic components of $f(t)$ and adding those responses at the output, as indicated in Figure 7.2. This concept is used throughout all of the engineering disciplines not only to simplify the solution but, more important, to *gain insight* into how the system processes the individual components of the input. Electrical engineers, for example, design electrical *filters* which are used in radios to filter out unwanted radio station transmissions.

### 7.1  PERIODIC FUNCTIONS

There are a large number of engineering waveforms that repeat themselves in blocks of a *period T*. These are said to be *periodic* and have the property that

$$f(t \pm kT) = f(t) \qquad \text{for } k = 1, 2, 3, \ldots \tag{7.4}$$

An example of a periodic function is shown in Figure 7.3. Periodic functions are said to have a *frequency*, in units of hertz (Hz), that is the *reciprocal* of the period: $f = 1/T$. The sine and cosine functions, $\sin 2\pi f$ and $\cos 2\pi f$, are periodic with period $T = 1/f$. For example, a 1-GHz ($10^9$-Hz) digital clock waveform in a personal computer has a period of 1 ns. Periodic functions can be represented by the decomposition shown in (7.1), and the individual functions $f_i(t)$ are periodic with frequencies that are integer multiples of the same basic frequency, $f = 1/T$, as the original function. The first term, $f_0$, in (7.1) is a constant and represents the *average value* of the function over one period:

$$f_0 = \frac{1}{T} \int_{t=0}^{T} f(t)\, dt \tag{7.5}$$

This means that we obtain *the area under the curve over one period and average that over a period.* In electrical engineering terminology, this would represent the dc component of the waveform. The Fourier series that we study in the next section uses the sinusoid as the basic function:

$$f(t) = f_0 + \sum_{n=1}^{\infty} (a_n \sin n\omega_o t + b_n \cos n\omega_o t) = f_0 + \sum_{n=1}^{\infty} c_n \cos(n\omega_o t + \theta_n) \tag{7.6}$$

Note that the sine and cosine terms have radian frequencies that are *integer multiples* of the basic radian frequency of the original periodic function: $\omega_o = 2\pi/T$. These summations require (theoretically) an infinite number of terms to represent the function. In practice, using only the first few terms will give an adequate representation of the original function.

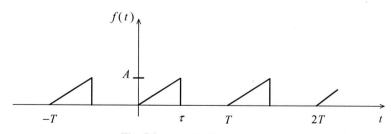

**Fig. 7.3.** A periodic waveform.

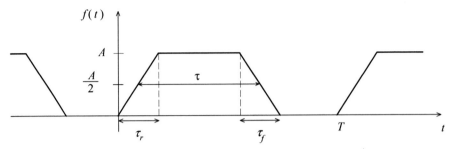

**Fig. 7.4.** A trapezoidal, digital clock waveform.

A very common periodic function is the clock waveform in a digital computer that is illustrated in Figure 7.4. The waveform has an amplitude $A$, a period $T$, a risetime $\tau_r$, and a falltime $\tau_f$. The *frequency* of the waveform is the reciprocal of the period, $f = 1/T$. The pulse width is denoted as $\tau$ and represents the time between the 50% levels of the waveform. A 100-MHz (period of $T = 10$ ns) digital clock waveform having $A = 5$ V, a risetime and a falltime of 1 ns, and a 50% duty cycle ($\tau = T/2$) is somewhat typical of digital waveforms. It can be decomposed, using the Fourier series that we study in the next section, into the sum of a constant and various sinusoids that have frequencies that are *integer multiples* of the basic clock frequency (harmonics) as

$$f(t) = 2.5 + 3.131\cos(\omega_o t - 108°) + 0.9108\cos(3\omega_o t - 144°)$$
$$+ 0.4053\cos(5\omega_o t - 180°) + 0.1673\cos(7\omega_o t - 216°)$$
$$+ 0.0387\cos(9\omega_o t + 108°) + \cdots$$

where the radian frequency is $\omega_o = 2\pi/T$. Because of the 50% duty cycle of the waveform, the even harmonics are absent for this waveform. Figure 7.5(a) compares the actual waveform and its reconstruction using only the first five terms of the Fourier series above. Since the even harmonic terms, $n = 2, 4, 6, 8, \ldots$ in the series above for this waveform are zero, only the sum of the constant term and the terms for $n = 1, 3, 5$ is plotted. Figure 7.5(b) compares the actual waveform and its reconstruction using only the first nine terms of the Fourier series above. Again, since the even harmonic terms, $n = 2, 4, 6, 8, \ldots$, in the series above for this waveform are zero, only the sum of the constant term and the terms for $n = 1, 3, 5, 7, 9$ is plotted in Figure 7.5(b). The individual components are also plotted in Figure 7.5(b), to show their contribution to the total sum. As can be seen, adding more terms gives a better approximation to the original waveform, which is shown in Figure 7.4.

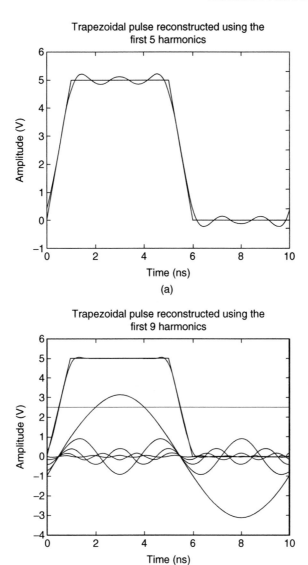

**Fig. 7.5.** Plot of 100-MHz clock (a) using five harmonics, and (b) using nine harmonics.

## 7.2  THE FOURIER SERIES

The Fourier series uses as the basic $f_n(t)$ functions the trigonometric sine and cosine functions, $\sin n\omega_o t$ and $\cos n\omega_o t$. Remember that the arguments of the sine and cosine functions must be in radians. The basic radian frequency of $\omega_o = 2\pi/T$ has units of radians per second. The Fourier series of a periodic function $f(t)$ has the form

$$f(t) = f_0 + \sum_{n=1}^{\infty} a_n \sin n\omega_o t + \sum_{n=1}^{\infty} b_n \cos n\omega_o t \tag{7.7}$$

where the radian frequency of the periodic waveform is

$$\omega_o = \frac{2\pi}{T} \tag{7.8}$$

The first term is again the average value of the function over one period.

$$f_0 = \frac{1}{T}\int_{t=0}^{T} f(t)\,dt \tag{7.9}$$

The remaining coefficients of the sine and cosine terms can be obtained using

$$a_n = \frac{2}{T}\int_0^T f(t)\sin n\omega_o t\,dt \tag{7.10a}$$

and

$$b_n = \frac{2}{T}\int_0^T f(t)\cos n\omega_o t\,dt \tag{7.10b}$$

With our understanding of basic trigonometry in Chapter 2, it is rather simple to show how these formulas for the coefficients in (7.9) and (7.10) come about. The formula for $f_0$ in (7.9) can be obtained by integrating both sides of the Fourier series in (7.7) over a period:

$$\int_0^T f(t)\,dt = f_0 \underbrace{\int_0^T dt}_{T} + \sum_{n=1}^{\infty} a_n \underbrace{\int_0^T \sin n\omega_o t\,dt}_{0} + \sum_{n=1}^{\infty} b_n \underbrace{\int_0^T \cos n\omega_o t\,dt}_{0}$$

$$= f_0 T \tag{7.11}$$

The integrals of a sine and a cosine over a period are zero since they have the same positive area as negative area over a period.

To obtain the formulas in (7.10) we first multiply both sides of the Fourier series in (7.7) by $\sin m\omega_o t$, and integrate that from 0 to $T$:

$$\int_0^T \sin m\omega_o t\, f(t)\, dt \;=\; f_0 \underbrace{\int_0^T \sin m\omega_o t\, dt}_{0} + \sum_{n=1}^{\infty} a_n \underbrace{\int_0^T \sin m\omega_o t \sin n\omega_o t\, dt}_{\substack{T/2 \;\; m=n \\ 0 \quad m\neq n}}$$

$$+ \sum_{n=1}^{\infty} b_n \underbrace{\int_0^T \sin m\omega_o t \cos n\omega_o t\, dt}_{0}$$

$$= a_n\, \frac{T}{2} \tag{7.12}$$

The first integral in (7.12) is zero since $f_0$ is a constant, and the integral of a sine over a period is zero: It has the same positive area as negative area over one period. The last integral in (7.12) is also zero. Using a trigonometric identity derived in Chapter 2:

$$\sin A \cos B = \frac{1}{2}\sin(A+B) + \frac{1}{2}\sin(A-B)$$

the integrand of the last term becomes the sum of two sines as

$$\sin m\omega_o t \cos n\omega_o t = \frac{1}{2}\sin[(m+n)\omega_o t] + \frac{1}{2}\sin[(m-n)\omega_o t]$$

whose integrals over a period are zero. The middle integral in (7.12) is $T/2$ for $m = n$ and zero for $m \neq n$. This is because of another trigonometric identity from Chapter 2:

$$\sin A \sin B = \frac{1}{2}\cos(A-B) - \frac{1}{2}\cos(A+B)$$

and the integrand of the middle term becomes the sum of two cosines:

$$\sin m\omega_o t \sin n\omega_o t = \frac{1}{2}\cos[(m-n)\omega_o t] - \frac{1}{2}\cos[(m+n)\omega_o t]$$

The integral over one period of each term is zero. But if $m = n$, the first term becomes $1/2$ whose integral over one period is $T/2$. Rearranging the result in (7.12) gives (7.10a). Equation (7.10b) follows in a very similar fashion: Multiply both sides of the Fourier series in (7.7) by $\cos m\omega_o t$, and integrate that from 0 to $T$ (which you should verify).

The Fourier series in (7.7) contains both sines and cosines but can be written in alternative forms containing only sines (with an angle) or only cosines (with an angle) as

$$f(t) = f_0 + \sum_{n=1}^{\infty} c_n \cos(n\omega_o t + \theta_n) \qquad (7.13a)$$

or

$$f(t) = f_0 + \sum_{n=1}^{\infty} c_n \sin(n\omega_o t + \theta_n + 90°) \qquad (7.13b)$$

We used the basic trigonometry result from Chapter 2, $\sin(\theta + 90°) = \cos\theta$ to convert (7.13a) to (7.13b). The alternative forms in (7.13) can be obtained by using the trigonometric identities derived in Chapter 2:

$$\cos(A + B) = \cos A \cos B - \sin A \sin B$$

$$\sin(A + B) = \sin A \cos B + \cos A \sin B$$

Write (7.13a) using the first identity as

$$f(t) = f_0 + \sum_{n=1}^{\infty} c_n \cos(n\omega_o t + \theta_n)$$

$$= f_0 - \sum_{n=1}^{\infty} c_n \sin\theta_n \sin n\omega_o t + \sum_{n=1}^{\infty} c_n \cos\theta_n \cos n\omega_o t$$

Comparing this to (7.7), we identify

$$a_n = -c_n \sin\theta_n \qquad (7.14a)$$

$$b_n = c_n \cos\theta_n \qquad (7.14b)$$

Using Euler's identity from Chapter 2, we can write this in complex form as

$$c_n \angle\theta_n = c_n(\cos\theta_n + j\sin\theta_n)$$
$$= c_n e^{j\theta_n} \qquad (7.15)$$
$$= b_n - ja_n$$

Substituting the formulas for $a_n$ and $b_n$ from (7.10) gives

$$c_n \angle\theta_n = b_n - ja_n$$

$$= \frac{2}{T} \int_0^T f(t) \underbrace{(\cos n\omega_o t - j\sin n\omega_o t)}_{e^{-jn\omega_o t}} dt$$

Hence we obtain a direct formula to obtain $c_n$ and $\theta_n$ as

$$\boxed{c_n \angle \theta_n = \frac{2}{T} \int_0^T f(t) e^{-jn\omega_0 t} \, dt}$$
(7.16)

---

### Example

Determine the Fourier series for the "square wave" periodic function shown in Figure 7.6.

From (7.9) we obtain $f_0$ as

$$f_0 = \frac{1}{T} \int_{t=0}^{T} f(t) \, dt$$

$$= \frac{1}{T} \int_{t=0}^{\tau} A \, dt + \frac{1}{T} \int_{\tau}^{T} 0 \, dt$$

$$= A \frac{\tau}{T}$$

This could have been more easily determined as

$$f_0 = \frac{\text{area}}{T} = \frac{A \times \tau}{T}$$

The remaining coefficients of the sine and cosine terms can be obtained using (7.10):

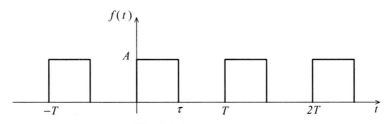

**Fig. 7.6.** A square wave.

$$a_n = \frac{2}{T} \int_0^\tau A \sin n\omega_o t \, dt$$

$$= \frac{2A}{T} \left( -\frac{1}{n\omega_o} \cos n\omega_o t \right)_{t=0}^{t=\tau}$$

$$= \frac{2A}{n\omega_o T} (1 - \cos n\omega_o \tau)$$

$$= \frac{A}{n\pi} \left[ 1 - \cos\left( 2n\pi \frac{\tau}{T} \right) \right]$$

and

$$b_n = \frac{2}{T} \int_0^\tau A \cos n\omega_o t \, dt$$

$$= \frac{2A}{T} \left( \frac{1}{n\omega_o} \sin n\omega_o t \right)_{t=0}^{t=\tau}$$

$$= \frac{2A}{n\omega_o T} (\sin n\omega_o \tau)$$

$$= \frac{A}{n\pi} \left[ \sin\left( 2n\pi \frac{\tau}{T} \right) \right]$$

where we have substituted (as it is always a good idea to do) the fundamental relationship

$$\boxed{n\omega_o = n\frac{2\pi}{T}}$$

The $c_n$ and $\theta_n$ in the alternative forms in (7.13) can be obtained from (7.15) or (7.16) as

$$c_n \angle \theta_n = \frac{2}{T} \int_0^\tau A e^{-jn\omega_o t} \, dt$$

$$= \frac{2A}{T} \left( -\frac{1}{jn\omega_o} e^{-jn\omega_o t} \right)_{t=0}^{t=\tau}$$

$$= j \frac{2A}{n\omega_o T} \left[ \underbrace{e^{-jn\omega_o \tau} - 1}_{\cos n\omega_o \tau - j \sin n\omega_o \tau - 1} \right]$$

$$= \frac{2A}{n\omega_o T} [\sin n\omega_o \tau + j(\cos n\omega_o \tau - 1)]$$

$$= \frac{A}{n\pi} \left[ \sin\left( 2n\pi \frac{\tau}{T} \right) + j\left( \cos\left( 2n\pi \frac{\tau}{T} \right) - 1 \right) \right]$$

and we have substituted $n\omega_o = n(2\pi/T)$ and used the fact that $1/j = -j$. Comparing this to the results for $a_n$ and $b_n$, we see that they satisfy (7.15), as they should.

If the duty cycle is 50%, $\tau = T/2$, or, in other words, the pulse is "on" for half the period and "off" for the other half of the period (which is typical for digital clock waveforms), the arguments of the sine and cosine terms simplify to

$$2n\pi \frac{\tau}{T} = n\pi \qquad \text{for } \tau = \frac{T}{2}$$

and these results simplify to

$$a_n = \frac{2A}{n\pi} \qquad \text{for } n \text{ odd}$$

$$a_n = 0 \qquad \text{for } n \text{ even}$$

$$b_n = 0 \qquad \text{for all } n$$

You should check these results. For a 50% duty cycle, the result for the $c_n \angle \theta_n$ simplifies to

$$c_n \angle \theta_n = \frac{A}{n\pi}[\sin n\pi + j(\cos n\pi - 1)]$$

$$= \begin{cases} -j\dfrac{2A}{n\pi} = \dfrac{2A}{n\pi}\angle -90° & n \text{ odd} \\ 0 & n \text{ even} \end{cases}$$

---

*Example*

Determine the Fourier series coefficients for the waveform in Figure 7.3.

To describe this waveform, we must write the equation of a straight line that was a fundamental skill covered in Chapter 2 (you should check this):

$$f(t) = \begin{cases} \dfrac{A}{\tau}t & 0 \leq t \leq \tau \\ 0 & \tau \leq t \leq T \end{cases}$$

The constant term, $f_0$, is obtained from (7.9) as

$$f_0 = \frac{1}{T}\frac{A}{\tau}\int_{t=0}^{\tau} t\,dt$$

$$= \frac{A}{T\tau}\left(\frac{t^2}{2}\right)_{t=0}^{t=\tau}$$

$$= \frac{A}{2}\frac{\tau}{T}$$

This could have been more easily obtained from

$$f_0 = \frac{\text{area}}{T} = \frac{(1/2) \times \tau \times A}{T}$$

The $a_n$ and $b_n$ coefficicients are obtained from (7.10):

$$a_n = \frac{2}{T} \int_0^T f(t) \sin n\omega_0 t \, dt$$

$$= \frac{2}{T} \frac{A}{\tau} \int_0^\tau t \sin n\omega_0 t \, dt$$

$$= \frac{2A}{T\tau} \frac{1}{(n\omega_0)^2} \int_0^{n\omega_0 \tau} \lambda \sin \lambda \, d\lambda$$

$$= \frac{2A}{T\tau} \frac{1}{(n\omega_0)^2} (\sin \lambda - \lambda \cos \lambda)_{\lambda=0}^{\lambda=n\omega_0 \tau}$$

$$= \frac{A}{2n^2 \pi^2} \frac{T}{\tau} \left[ \sin\left(2n\pi \frac{\tau}{T}\right) - \left(2n\pi \frac{\tau}{T}\right) \cos\left(2n\pi \frac{\tau}{T}\right) \right]$$

In evaluating this integral we have used a change of variables $\lambda = n\omega_0 t$ (Chapter 2). Hence we (1) replace $t = (1/n\omega_0)\lambda$ and $dt = (1/n\omega_0)d\lambda$, and (2) replace the limits: $t = 0 \rightarrow \lambda = 0$ and $t = \tau \rightarrow \lambda = n\omega_0 \tau$. We also substituted the basic relationship $n\omega_0 = 2n\pi/T$, and have "looked up," in a table of integrals, the integral

$$\int \lambda \sin \lambda \, d\lambda = \sin \lambda - \lambda \cos \lambda$$

The coefficient $b_n$ is determined similarly from (7.10b) as

$$b_n = \frac{2}{T} \int_0^T f(t) \cos n\omega_0 t \, dt$$

$$= \frac{2}{T} \frac{A}{\tau} \int_0^\tau t \cos n\omega_0 t \, dt$$

$$= \frac{2A}{T\tau} \frac{1}{(n\omega_0)^2} \int_0^{n\omega_0 \tau} \lambda \cos \lambda \, d\lambda$$

$$= \frac{2A}{T\tau} \frac{1}{(n\omega_0)^2} (\cos \lambda + \lambda \sin \lambda)_{\lambda=0}^{\lambda=n\omega_0 \tau}$$

$$= \frac{A}{2n^2 \pi^2} \frac{T}{\tau} \left[ \cos\left(2n\pi \frac{\tau}{T}\right) + \left(2n\pi \frac{\tau}{T}\right) \sin\left(2n\pi \frac{\tau}{T}\right) - 1 \right]$$

and we have "looked up," in a table of integrals, the integral

$$\int \lambda \cos \lambda \, d\lambda = \cos \lambda + \lambda \sin \lambda$$

As a "sanity check," you should differentiate both integral identities to show that they give the integrands. For example,

$$\frac{d}{d\lambda}(\cos \lambda + \lambda \sin \lambda) = -\sin \lambda + \sin \lambda + \lambda \cos \lambda = \lambda \cos \lambda$$

where we used the chain rule on the second part. Similarly,

$$\frac{d}{d\lambda}(\sin \lambda - \lambda \cos \lambda) = \cos \lambda - \cos \lambda + \lambda \sin \lambda = \lambda \sin \lambda$$

---

## 7.3  THE FOURIER TRANSFORM

There are many situations where a waveform consists of a *single pulse* that does not repeat periodically in time. For example, suppose that we let the period $T$ of the periodic waveform in Figure 7.3 go to infinity, $T \to \infty$. We are left with only one, nonrecurring pulse, as shown in Figure 7.7. How do we decompose this waveform into its sinusoidal components? The simple answer is to let $T$ go to infinity, $T \to \infty$, in the Fourier series representation for the corresponding periodic waveform in Figure 7.3 and see what this gives. For a periodic waveform, the frequency components occur only at integer multiples of the basic repetition frequency, $n\omega_0 = n(2\pi/T)$ for $n = 1, 2, 3, \ldots$ , as shown in Figure 7.8. The adjacent frequencies are separated by $\Delta\omega = \omega_0 = 2\pi/T$. There are no frequency components between these harmonic frequencies. But as we let $T \to \infty$, we observe the important result that *the discrete frequency components merge into a smooth continuum of frequencies*, as illustrated in Figure 7.8. So as $T \to \infty$, we replace $n\omega_0 \to \omega = 2\pi f$, and the pulse can be thought of as being composed of a continuum of sinusoidal components. This results in what is called the *Fourier transform*, which applies to *nonperiodic functions*.

Multiplying and dividing each summation in (7.7) by the separation between each of the harmonics, $\Delta\omega = \omega_0 = 2\pi/T$ gives

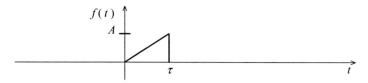

**Fig. 7.7.** A single pulse leading to the Fourier transform as $T \to \infty$.

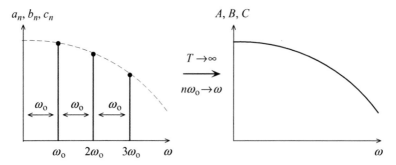

**Fig. 7.8.** Obtaining the Fourier transform from the Fourier series as $T \to \infty$.

$$f(t) = f_0 + \frac{1}{\omega_0} \sum_{n=1}^{\infty} a_n \sin n\omega_0 t \, \Delta\omega + \frac{1}{\omega_0} \sum_{n=1}^{\infty} b_n \cos n\omega_0 t \, \Delta\omega$$

$$= f_0 + \sum_{n=1}^{\infty} \frac{a_n}{\omega_0} \sin n\omega_0 t \, \Delta\omega + \sum_{n=1}^{\infty} \frac{b_n}{\omega_0} \cos n\omega_0 t \, \Delta\omega$$

so that the expansion coefficients in (7.10) now become, for this representation,

$$\frac{a_n}{\omega_0} = \frac{T}{2\pi} a_n = \frac{1}{\pi} \int_0^T f(t) \sin n\omega_0 t \, dt$$

$$\frac{b_n}{\omega_0} = \frac{T}{2\pi} b_n = \frac{1}{\pi} \int_0^T f(t) \cos n\omega_0 t \, dt$$

Replacing the summations with an integral in the limit as $T \to \infty$, and replacing $n\omega_0 \to \omega = 2\pi f$ and $\Delta\omega \to d\omega = 2\pi df$, gives for the *nonperiodic* function $f(t)$ such as is shown in Figure 7.7:

$$f(t) = \int_0^{\infty} \left[ 2\int_0^{\infty} f(t)\sin \omega t \, dt \right] \sin 2\pi f t \, df + \int_0^{\infty} \left[ 2\int_0^{\infty} f(t)\cos \omega t \, dt \right] \cos 2\pi f t \, df$$

Hence, the general form of the Fourier transform for a nonperiodic $f(t)$ becomes

$$\boxed{f(t) = \int_0^{\infty} A(\omega)\sin 2\pi f t \, df + \int_0^{\infty} B(\omega)\cos 2\pi f t \, df} \qquad (7.17)$$

where the coefficients $A(\omega)$ and $B(\omega)$ are

$$\boxed{A(\omega) = 2\int_0^{\infty} f(t)\sin \omega t \, dt} \qquad (7.18a)$$

$$\boxed{B(\omega) = 2\int_0^\infty f(t)\cos\omega t\,dt}$$ (7.18b)

Note the similarity with the equations for the coefficients in the Fourier series in (7.10).

As was the case for the Fourier series, these results can be written in terms of only a cosine having an angle:

$$\boxed{f(t) = \int_0^\infty C\cos(2\pi ft + \theta_C)\,df}$$ (7.19)

Substituting the trigonometric identity from Chapter 2 of

$$\cos(A+B) = \cos A\cos B - \sin A\sin B$$

into (7.19) gives

$$f(t) = \int_0^\infty C\cos(2\pi ft + \theta_C)\,df$$
$$= -\int_0^\infty C\sin\theta_C\,\sin 2\pi ft\,df + \int_0^\infty C\cos\theta_C\,\cos 2\pi ft\,df$$

Comparing this to (7.17) shows that

$$A(\omega) = -C\sin\theta_C$$
$$B(\omega) = C\cos\theta_C$$

Hence, the complex-valued coefficient

$$\begin{aligned}\mathbf{C}(\omega) &= C\angle\theta_C \\ &= Ce^{j\theta_C} \\ &= C\cos\theta_C + jC\sin\theta_C \\ &= B(\omega) - jA(\omega)\end{aligned}$$

is obtained by substituting (7.18) to give

$$\boxed{\begin{aligned}C\angle\theta_C &= B(\omega) - jA(\omega) \\ &= 2\int_0^\infty f(t)\underbrace{(\cos\omega t - j\sin\omega t)}_{e^{-j\omega t}}\,dt \\ &= 2\int_0^\infty f(t)e^{-j\omega t}\,dt\end{aligned}}$$ (7.20)

## *Example*

Determine the Fourier transform coefficients for the single rectangular pulse shown in Figure 7.9.
   From (7.18) we obtain

$$A(\omega) = 2\int_0^\infty f(t)\sin \omega t \, dt$$
$$= 2\int_0^\tau A\sin \omega t \, dt$$
$$= \frac{2A}{\omega}(-\cos \omega t)_{t=0}^{t=\tau}$$
$$= \frac{2A}{\omega}(1 - \cos \omega \tau)$$

and

$$B(\omega) = 2\int_0^\infty f(t)\cos \omega t \, dt$$
$$= 2\int_0^\tau A\cos \omega t \, dt$$
$$= \frac{2A}{\omega}(\sin \omega t)_{t=0}^{t=\tau}$$
$$= \frac{2A}{\omega}(\sin \omega \tau)$$

These should be compared with the Fourier series coefficients for the periodic waveform:

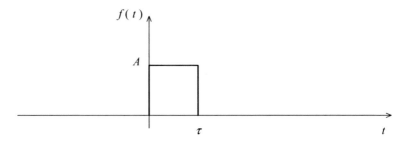

**Fig. 7.9.** A single rectangular pulse waveform.

$$a_n = \frac{2A}{n\omega_o T}(1 - \cos n\omega_o \tau) \quad \text{and} \quad b_n = \frac{2A}{n\omega_o T}(\sin n\omega_o \tau)$$

Observe that

$$A(\omega) = Ta_n(n\omega_o \to \omega) \quad \text{and} \quad B(\omega) = Tb_n(n\omega_o \to \omega)$$

as can be seen by comparing (7.10) and (7.18). Once the Fourier series coefficients are obtained for the periodic function, the Fourier transform coefficients for a single pulse, represented by the basic waveform over one period, may be obtained from them by (1) multiplying $a_n$ and $b_n$ by $T$ and (2) replacing $n\omega_o \to \omega$. This is a general result for any waveform.

Similarly, the alternative representation in (7.19) can be used with $C(\omega)$ obtained from (7.20):

$$C\angle\theta_C = 2\int_0^\infty f(t)e^{-j\omega t}\, dt$$
$$= 2\int_0^\tau Ae^{-j\omega t}\, dt$$
$$= \frac{2A}{-j\omega}(e^{-j\omega\tau} - 1)$$
$$= j\frac{2A}{\omega}(e^{-j\omega\tau} - 1)$$
$$= \frac{2A}{\omega}[\sin\omega\tau + j(\cos\omega\tau - 1)]$$

where we used the complex algebra results of $1/j = -j$ and $je^{-j\omega t} = j\cos\omega t + \sin\omega t$. Compare this to the corresponding result for the periodic waveform obtained in a previous example:

$$C_n\angle\theta_n = \frac{2A}{n\omega_o T}[\sin n\omega_o t + j(\cos n\omega_o t - 1)]$$

Do you see how to obtain the result for a single pulse from the result for a periodic pulse train?

# 8 The Laplace Transform

In Chapter 7 we illustrated the idea of using a *transform* to solve mathematical problems. Alternatively, the Fourier series or the Fourier transform represents a function of time in the *time domain* as a sum of sinusoidal functions, which is said to be in the *frequency domain*. Using superposition for a linear system, the response of the system to these individual sinusoidal components can be determined, and then the response of the system to the original time-domain function can be determined as the sum of these individual frequency-domain responses. Another example of solving a problem by transforming to another domain where the mathematics is easier is the logarithm. Multiplication of numbers is more difficult than their addition or subtraction. If we wish to multiply two numbers as $z = xy$, we could take the logarithm (to any base) of this to give $\log_b z = \log_b x + \log_b y$ and have transformed a product into a sum. We can then return to the original domain by taking the antilog, $z = b^{\log_b z} = b^{\log_b x + \log_b y}$, to obtain the desired result.

The Laplace transform is a similar idea for solving differential equations. Solving an ordinary differential equation directly, as in Chapter 4, requires that we (1) solve the homogeneous differential equation, (2) solve the nonhomogeneous differential equation, and (3) apply the initial conditions to the sum of these solutions to determine the constants in the homogeneous solution. The Laplace transform method changes the differential equation into an algebraic equation whose manipulation is much simpler. With the Laplace transform method we just "turn the crank," so to speak, and the solution to the differential equation "comes out."

Throughout this chapter we represent the variables as being functions of time, which we again represent as $t$. However, you should be flexible and adapt this technique readily to variables that are a function of other types of independent variables [e.g., $f(x)$, with $x$ representing position]. The Laplace transform of a function $f(t)$ is defined by

$$\boxed{\begin{aligned} F(s) &\equiv \int_0^\infty f(t) e^{-st} dt \\ F(s) &\Leftrightarrow f(t) \end{aligned}} \tag{8.1}$$

We consistently denote the Laplace transform as a capital letter and show it as a function of the Laplace transform variable $s$ (which is a complex-valued variable):

*Essential Math Skills for Engineers*, By Clayton R. Paul
Copyright © 2009 John Wiley & Sons, Inc.

$f(t) \Leftrightarrow F(s)$. Since we want to use this to solve differential equations whose responses (and initial conditions) start at $t = 0$, the lower limit of the Laplace transform is $t = 0$. We transform the time-domain functions into the Laplace or $s$-domain, manipulate them to achieve a solution in that domain, and then transform that result back to the time domain to achieve the solution desired.

## 8.1   TRANSFORMS OF IMPORTANT FUNCTIONS

There are four basic functions of time that occur most frequently throughout engineering analyses: (1) the constant $f(t) = K$, (2) a linear function of time, $f(t) = Kt$, (3) the exponential function, $f(t) = Ke^{at}$, and (4) the sinusoidal function, $f(t) = K\sin(\omega t + \theta)$ or $f(t) = K\cos(\omega t + \theta)$. Once you have mastered the Laplace transform for these functions, you can readily extend your knowledge to other functions. We henceforth use the term "transform" to refer to the "Laplace transform."

The transform of the constant is

$$\boxed{K \Leftrightarrow \frac{K}{s}} \tag{8.2}$$

This can be shown simply by substituting into (8.1) and carrying out the integration:

$$F(s) = \int_0^\infty Ke^{-st} \, dt$$

$$= -\frac{K}{s} e^{-st} \Big|_{t=0}^{t \to \infty}$$

$$= \frac{K}{s}$$

The transform of a linear function of $t$ is

$$\boxed{Kt \Leftrightarrow \frac{K}{s^2}} \tag{8.3}$$

This can also be shown by substituting into (8.1) and carrying out the integration:

$$F(s) = \int_0^\infty Kte^{-st} \, dt$$

$$= K\underbrace{\left(-\frac{1}{s} te^{-st}\right)\Big|_{t=0}^{t \to \infty}}_{0} + \frac{K}{s} \int_0^\infty e^{-st} \, dt$$

$$= -\frac{K}{s^2} \left(e^{-st}\right)_{t=0}^{t \to \infty}$$

$$= \frac{K}{s^2}$$

This result was obtained by integrating by parts: $\int u\,dv = uv - \int v\,du$ where $u = t$ and $dv = e^{-st}\,dt$, so that $du = 1\,dt$ and $v = -e^{-st}/s$. Alternatively, we could have "looked up" the integral in a table of integrals:

$$\int xe^{ax}\,dx = e^{ax}\left(\frac{x}{a} - \frac{1}{a^2}\right)$$

In evaluating the first part of this we have used the important fact that $\lim_{t\to\infty} te^{-st} = 0$, since as $t$ goes to $\infty$, $e^{-st}$ goes to 0 faster than $t$ goes to $\infty$. To show this, use l'Hôpital's rule from Chapter 2. Differentiate the numerator and the denominator and then let $t \to \infty$: $te^{-st} = t/e^{st} \to 1/se^{st} \to 0$. You can also show, by integration by parts, a general result:

$$\boxed{Kt^n \Leftrightarrow n!\,\frac{K}{s^{n+1}}} \tag{8.4}$$

where $n!$ denotes $n$ factorial: $n! = 1 \times 2 \times 3 \times \dots \times n$. You should verify this by using integration by parts repeatedly for $n = 1, 2, 3, \dots$ to see a pattern.

The transform of the exponential function is

$$\boxed{Ke^{at} \Leftrightarrow \frac{K}{s-a}} \tag{8.5}$$

This can be shown by again simply substituting into (8.1) and carrying out the integration:

$$F(s) = \int_0^\infty Ke^{at}e^{-st}\,dt$$

$$= \int_0^\infty Ke^{-(s-a)t}\,dt$$

$$= -\frac{K}{s-a}e^{-(s-a)t}\Big|_{t=0}^{t\to\infty}$$

$$= \frac{K}{s-a}$$

The transforms of $f(t) = K\sin(\omega t + \theta)$ or $f(t) = K\cos(\omega t + \theta)$ can also be obtained by substitution into (8.1) and "looking up" the resulting integral in a table of integrals. A much simpler way to evaluate the resulting integral is to substitute Euler's identity (Chapter 2) for the sine function:

$$\sin(\omega t + \theta) = \frac{e^{j(\omega t + \theta)} - e^{-j(\omega t + \theta)}}{2j}$$

to give

$$F(s) = \int_0^\infty K \sin(\omega t + \theta) e^{-st}\, dt$$

$$= K \int_0^\infty \frac{e^{j(\omega t+\theta)} - e^{-j(\omega t+\theta)}}{2j} e^{-st}\, dt$$

$$= \frac{K}{2j} \int_0^\infty e^{j(\omega t+\theta)} e^{-st}\, dt - \frac{K}{2j} \int_0^\infty e^{-j(\omega t+\theta)} e^{-st}\, dt$$

$$= \frac{K}{2j} \int_0^\infty e^{-(s-j\omega)t} e^{j\theta}\, dt - \frac{K}{2j} \int_0^\infty e^{-(s+j\omega)t} e^{-j\theta}\, dt$$

$$= \frac{K}{2j} \left[ -\frac{e^{j\theta}}{s-j\omega} e^{-(s-j\omega)t} + \frac{e^{-j\theta}}{s+j\omega} e^{-(s+j\omega)t} \right]_{t=0}^{t\to\infty}$$

$$= \frac{K}{2j} \left( \frac{e^{j\theta}}{s-j\omega} - \frac{e^{-j\theta}}{s+j\omega} \right)$$

$$= \frac{K}{2j} \frac{(s+j\omega)e^{j\theta} - (s-j\omega)e^{-j\theta}}{s^2+\omega^2}$$

$$= \frac{K}{s^2+\omega^2} \left( s\frac{e^{j\theta} - e^{-j\theta}}{2j} + \omega\frac{e^{j\theta} + e^{-j\theta}}{2} \right)$$

$$= K \frac{s\sin(\theta) + \omega\cos\theta}{s^2+\omega^2}$$

(*Note:* The detail makes this seem like a difficult integration, but it is not. It is very straightforward. I have simply written out every detail to show you the process.) Hence,

$$\boxed{K \sin(\omega t + \theta) \Leftrightarrow K \frac{s\sin\theta + \omega\cos\theta}{s^2+\omega^2}} \qquad (8.6a)$$

Using the trigonometric identity (Chapter 2), $\cos(\omega t + \theta) = \sin(\omega t + \theta + 90°)$, we obtain from this

$$K \cos(\omega t + \theta) \Leftrightarrow K \frac{s\sin(\theta+90°) + \omega\cos(\theta+90°)}{s^2+\omega^2}$$

Since $\sin(\theta + 90°) = \cos\theta$ and $\cos(\theta + 90°) = -\sin\theta$, we obtain

$$\boxed{K \cos(\omega t + \theta) \Leftrightarrow K \frac{s\cos\theta - \omega\sin\theta}{s^2+\omega^2}} \qquad (8.6b)$$

For $\theta = 0°$ these simplify to

$$\boxed{K \sin\omega t \Leftrightarrow K \frac{\omega}{s^2+\omega^2}} \qquad (8.6c)$$

$$\boxed{K\cos(\omega t) \Leftrightarrow K\frac{s}{s^2+\omega^2}}$$  (8.6d)

## 8.2  USEFUL TRANSFORM PROPERTIES

There are four very useful properties of the Laplace transform that can be used to obtain other transforms without carrying out the direct integration in (8.1). The first is the property of *linearity* of the Laplace transform:

$$\boxed{Af_1(t)+Bf_2(t) \Leftrightarrow AF_1(s)+BF_2(s)}$$  (8.7)

In other words, if we can decompose a function into the sum of two other functions *for which we already know the transform*, then the transform of the sum is the sum of the transforms. This property is a rather obvious result of the Laplace transform being an integral, which is a linear operation:

$$\int_0^\infty [Af_1(t)+Bf_2(t)]e^{-st}\,dt = A\underbrace{\int_0^\infty f_1(t)e^{-st}\,dt}_{F_1(s)} + B\underbrace{\int_0^\infty f_2(t)e^{-st}\,dt}_{F_2(s)}$$

The second is the property of *time shift*:

$$\boxed{f(t-t_0) \Leftrightarrow e^{-st_0}F(s)}$$  (8.8)

In other words, the transform of a function shifted to the right by $t_0$ is the transform of the *unshifted function* multiplied by $e^{-st_0}$. This can be shown by substituting into (8.1):

$$\int_0^\infty f(t-t_0)e^{-st}\,dt = e^{-st_0}\int_{-t_0}^\infty f(\lambda)e^{-s\lambda}\,d\lambda$$

$$= e^{-st_0}\underbrace{\int_{-t_0}^0 f(\lambda)e^{-s\lambda}\,d\lambda}_{0} + e^{-st_0}\underbrace{\int_0^\infty f(\lambda)e^{-s\lambda}\,d\lambda}_{F(s)}$$

where we have made a change of variables: $\lambda = t - t_0$, $d\lambda = dt$, and $t = 0 \rightarrow \lambda = -t_0$, $t = \infty \rightarrow \lambda = \infty$, and $f(t) = 0$ for $t < 0$.

---

*Example*

Determine the transform of the function shown in Figure 8.1.
  Observe that the original function can be decomposed into the sum of two "ramp functions," one of which is negated and shifted to the right by $\tau$:

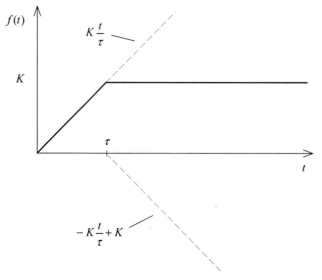

**Fig. 8.1**

$$f(t) = Kf_1(t) + Kf_2(t)$$

where

$$f_1(t) = \begin{cases} 0 & t < 0 \\ \dfrac{t}{\tau} & t > 0 \end{cases}$$

$$f_2(t) = \begin{cases} 0 & t < \tau \\ -\dfrac{t}{\tau} + 1 & t > \tau \end{cases}$$

(You should verify this.) Direct evaluation of (8.1) gives

$$F(s) = K\int_0^\infty \frac{t}{\tau} e^{-st}\, dt + K\int_\tau^\infty \left(-\frac{t}{\tau}+1\right) e^{-st}\, dt$$

$$= \frac{K}{\tau}\int_0^\infty t e^{-st}\, dt - \frac{K}{\tau}\int_\tau^\infty t e^{-st}\, dt + K\int_\tau^\infty e^{-st}\, dt$$

$$= \frac{K}{\tau s^2} - \frac{K}{\tau s^2} e^{-s\tau}$$

which you should verify. These are unnecessarily tedious integrals to evaluate. A much simpler and more immediate way of obtaining the answer *without evaluating any integrals* is by using the two properties above and the transform in (8.3) and recognizing that the total is the sum of a ramp function and that same ramp function negated and shifted to the right by $\tau$:

$$
\begin{aligned}
F(s) &= KF_1(s) + KF_2(s) \\
&= KF_1(s) - KF_1(s)e^{-s\tau} \\
&= \frac{K}{\tau s^2} - \frac{K}{\tau s^2}e^{-s\tau}
\end{aligned}
$$

The third property is that of *s-shift*:

$$
\boxed{e^{-at}f(t) \Leftrightarrow F(s+a)} \tag{8.9}
$$

In engineering problems we frequently encounter functions whose amplitudes decay at an exponential rate. If we know the transform of a function $f(t)$, the transform of that function, which is multiplied by the decaying exponential, $e^{-at}$, is simply the transform of the original function with $s$ replaced by $s + a$. This can be proven by direct substitution into (8.1):

$$
\int_0^\infty e^{-at}f(t)e^{-st}\,dt = \underbrace{\int_0^\infty f(t)e^{-(s+a)t}\,dt}_{F(s+a)}
$$

The final important property is the transform of a derivative. Substituting into (8.1) gives

$$
F(s) = \int_0^\infty \frac{df(t)}{dt}e^{-st}\,dt
$$

Using the property of integration by parts from Chapter 2 of $\int u\,dv = uv - \int v\,du$, we identify $u = e^{-st}$ and $dv = df(t)$, to give

$$
\begin{aligned}
F(s) &= \int_0^\infty \frac{df(t)}{dt}e^{-st}\,dt \\
&= [f(t)e^{-st}]_{t=0}^{t\to\infty} - \int_0^\infty(-se^{-st})f(t)\,dt \\
&= -f(0) + s\underbrace{\int_0^\infty f(t)e^{-st}\,dt}_{F(s)}
\end{aligned}
$$

Hence, the transform of the first derivative is

$$
\boxed{\frac{df(t)}{dt} \Leftrightarrow sF(s) - f(0)} \tag{8.10}
$$

where $f(t) \Leftrightarrow F(s)$ and $f(0)$ is the initial condition. The transform of the second derivative can be obtained by being clever and defining $g(t) = df(t)/dt$. Hence,

$$\frac{d^2 f(t)}{dt^2} = \frac{dg(t)}{dt} \Leftrightarrow sG(s) - g(0)$$

so that the transform of the second derivative is

$$\frac{d^2 f(t)}{dt^2} \Leftrightarrow s[sF(s) - f(0)] - \dot{f}(0) \tag{8.11}$$

$$= s^2 F(s) - s f(0) - \dot{f}(0)$$

and $\dot{f}(0) = \dfrac{df(t)}{dt}\bigg|_{t=0}$ is an initial condition. The transform of higher-order derivatives can be obtained easily by continuing in like fashion:

$$\frac{d^3 f(t)}{dt^3} \Leftrightarrow s\left[s^2 F(s) - sf(0) - \dot{f}(0)\right] - \ddot{f}(0) \tag{8.12}$$

$$= s^3 F(s) - s^2 f(0) - s\dot{f}(0) - \ddot{f}(0)$$

## 8.3   TRANSFORMING DIFFERENTIAL EQUATIONS

We are now ready to solve differential equations by transforming them. We use two examples from Chapter 4 of first- and second-order ordinary differential equations and a partial differential equation from Chapter 6.

---

### Example

Transform the first-order differential equation

$$\frac{dx(t)}{dt} + 5x(t) = 10\cos 2t \qquad x(0) = 1$$

This transforms to

$$[sX(s) - 1] + 5X(s) = \frac{10s}{s^2 + 4}$$

Solving this for the transform of $x(t)$ gives

$$X(s) = \frac{s^2 + 10s + 4}{(s+5)(s^2+4)}$$

$$= \frac{s^2 + 10s + 4}{s^3 + 5s^2 + 4s + 20}$$

and we have written this as a ratio of polynomials in $s$. You should verify this.

---

### Example

Transform the second-order differential equation

$$\frac{d^2x(t)}{dt^2} + 3\frac{dx(t)}{dt} + 2x(t) = 10 \qquad x(0) = 2, \quad \dot{x}(0) = 1$$

This transforms to

$$[s^2 X(s) - 2s - 1] + 3[sX(s) - 2] + 2X(s) = \frac{10}{s}$$

Solving this for the transform of $x(t)$ gives

$$X(s) = \frac{2s^2 + 7s + 10}{s(s^2 + 3s + 2)}$$

$$= \frac{2s^2 + 7s + 10}{s^3 + 3s^2 + 2s}$$

and we have written this as a ratio of polynomials in $s$. You should verify this.

---

### Example

The Laplace transform can also be used to simplify the solution of partial differential equations (Chapter 6). For example, the "wave equation" is

$$\frac{\partial^2 u(x,t)}{\partial^2 x} = \frac{1}{v^2}\frac{\partial^2 u(x,t)}{\partial^2 t}$$

The initial conditions in time are assumed to be zero, $u(x,0) = 0$ and $\dot{u}(x,0) = 0$. Taking the Laplace transform of both sides with respect to $t$ removes the time variable and gives an ordinary differential equation that is only a function of $x$ ($s$ is a constant now):

$$\frac{d^2 U(x, s)}{dx^2} - \frac{s^2}{v^2} U(x, s) = 0$$

Since this transformed equation is homogeneous, its total solution is just the homogeneous solution. The characteristic equation is

$$p^2 - \frac{s^2}{v^2} = \left(p + \frac{s}{v}\right)\left(p - \frac{s}{v}\right)$$

whose roots are $p_1 = -s/v$ and $p_2 = s/v$. The solution to the transformed ordinary differential equation is (Chapter 4)

$$U(x, s) = C_1(s)e^{-sx/v} + C_2(s)e^{sx/v}$$

Using the time-shift property in (8.8) gives the solution as

$$u(x, t) = c_1\left(t - \frac{x}{v}\right) + c_2\left(t + \frac{x}{v}\right)$$

where $c_1(t) \Leftrightarrow C_1(s)$ and $c_2(t) \Leftrightarrow C_2(s)$. This is precisely the form of the solution to the wave equation as the sum of two "traveling waves" that was obtained rather laboriously in equation (6.19).

---

### 8.4   OBTAINING THE INVERSE TRANSFORM USING PARTIAL FRACTION EXPANSIONS

The only task remaining is to obtain the inverse Laplace transform of the results in Section 8.3. We obtain these, once again, by being clever. The basic idea is to *write the transformed solution as the sum of several functions of s for which we already know the inverse transform.* For example, the common ones have been obtained previously and are

$$Ke^{-at} \Leftrightarrow \frac{K}{s + a}$$

$$Kte^{-at} \Leftrightarrow \frac{K}{(s + a)^2}$$

As shown by the examples in Section 8.3, the Laplace transforms of the desired solution are in the form of a *ratio of polynomials in s*:

$$F(s) = M \frac{s^m + b_1 s^{m-1} + \cdots + b_{m-1} s + b_m}{s^n + a_1 s^{n-1} + \cdots + a_{n-1} s + a_n} \qquad (8.13)$$

The numerator polynomial is of order $m$, and the denominator polynomial is of order $n$ and $m < n$. The key is to expand this into the sum of functions of $s$ for which we immediately know the inverse transform. To do so, we first factor the denominator polynomial into its $n$ factors as

$$s^n + a_1 s^{n-1} + \cdots + a_{n-1} s + a_n = (s + p_1)(s + p_2) \cdots (s + p_n) \qquad (8.14)$$

where the $n$ roots of it are $-p_1, -p_2, \ldots, -p_n$. Then we can write (8.13) as

$$
\begin{aligned}
F(s) &= M \frac{s^m + b_1 s^{m-1} + \cdots + b_{m-1} s + b_m}{s^n + a_1 s^{n-1} + \cdots + a_{n-1} s + a_n} \\
&= M \frac{s^m + b_1 s^{m-1} + \cdots + b_{m-1} s + b_m}{(s + p_1)(s + p_2) \cdots (s + p_n)} \\
&= \frac{M_1}{s + p_1} + \frac{M_2}{s + p_2} + \cdots + \frac{M_n}{s + p_n}
\end{aligned}
\qquad (8.15)
$$

If we can determine the constants $M_1, M_2, \ldots, M_n$, the inverse transform is immediately seen to be

$$f(t) = M_1 e^{-p_1 t} + M_2 e^{-p_2 t} + \cdots + M_n e^{-p_n t} \qquad (8.16)$$

This is the key idea. The process of expanding (8.15) into the sum of the elementary factors is called the method of *partial fraction expansion*, which we address next.

The roots of the denominator polynomial in (8.14) can be only one of three forms: (1) real and distinct, (2), real and repeated, or (3) complex conjugate (possibly repeated for $n > 2$). To obtain the coefficients $M_1, M_2, \ldots, M_n$ in (8.15), we use the "cover-up rule." To obtain $M_i$, we multiply $F(s)$ by the denominator factor corresponding to it, $(s + p_i)$; that is, we "cover it up" in the denominator of $F(s)$:

$$
\begin{aligned}
(s + p_i) F(s) &= (s + p_i) M \frac{s^m + b_1 s^{m-1} + \cdots + b_{m-1} s + b_m}{(s + p_1) \cdots (s + p_i) \cdots (s + p_n)} \\
&= M \frac{s^m + b_1 s^{m-1} + \cdots + b_{m-1} s + b_m}{(s + p_1) \cdots (s + p_{i-1})(s + p_{i+1}) \cdots (s + p_n)} \\
&= (s + p_i) \frac{M_1}{s + p_1} + \cdots + M_i + \cdots + (s + p_i) \frac{M_n}{s + p_n}
\end{aligned}
\qquad (8.17)
$$

We then evaluate what remains by substituting $s = -p_i$, leaving

$$\boxed{M_i = [(s + p_i)F(s)]_{s=-p_i}} \tag{8.18}$$

---

### Example

For example, suppose that

$$F(s) = \frac{3s + 4}{s(s + 1)(s + 2)}$$

The denominator roots are $s = 0$, $s = -1$, and $s = -2$. So the partial fraction expansion would be of the form (always write out the anticipated form of the expansion)

$$F(s) = \frac{3s + 4}{s(s + 1)(s + 2)}$$
$$= \frac{M_1}{s} + \frac{M_2}{s + 1} + \frac{M_3}{s + 2}$$

With the cover-up rule, we obtain

$$M_1 = [sF(s)]_{s=0}$$
$$= \left[\frac{3s + 4}{(s + 1)(s + 2)}\right]_{s=0}$$
$$= \frac{4}{1 \cdot 2}$$
$$= 2$$

$$M_2 = [(s + 1)F(s)]_{s=-1}$$
$$= \left[\frac{3s + 4}{s(s + 2)}\right]_{s=-1}$$
$$= \frac{1}{-1 \cdot 1}$$
$$= -1$$

$$M_3 = [(s + 2)F(s)]_{s=-2}$$
$$= \left[\frac{3s + 4}{s(s + 1)}\right]_{s=-2}$$
$$= \frac{-2}{-2 \cdot -1}$$
$$= -1$$

Therefore, the partial fraction expansion is

$$F(s) = \frac{2}{s} - \frac{1}{s+1} - \frac{1}{s+2}$$

You should always "sanity check," this by multiplying it out to see if any mistakes were made in the partial fraction expansion:

$$
\begin{aligned}
F(s) &= \frac{2}{s} - \frac{1}{s+1} - \frac{1}{s+2} \\
&= \frac{2(s+1)(s+2) - 1s(s+2) - 1s(s+1)}{s(s+1)(s+2)} \\
&= \frac{2(s^2 + 3s + 2) - 1(s^2 + 2s) - 1(s^2 + s)}{s(s+1)(s+2)} \\
&= \frac{3s+4}{s(s+1)(s+2)}
\end{aligned}
$$

The inverse transform of this is easily seen to be

$$f(t) = 2 - e^{-t} - e^{-2t}$$

---

### Example

Solve the first-order differential equation

$$\frac{dx(t)}{dt} + 5x(t) = 10\cos 2t \qquad x(0) = 1$$

The transform of this is

$$[sX(s) - 1] + 5X(s) = \frac{10s}{s^2 + 4}$$

Solving for $X(s)$ gives, as was obtained in a previous example,

$$
\begin{aligned}
X(s) &= \frac{s^2 + 10s + 4}{(s+5)(s^2 + 4)} \\
&= \frac{s^2 + 10s + 4}{s^3 + 5s^2 + 4s + 20}
\end{aligned}
$$

Factoring the denominator gives

$$X(s) = \frac{s^2 + 10s + 4}{(s+5)(s+j2)(s-j2)}$$

$$= \frac{M_1}{s+5} + \frac{M_2}{s+j2} + \frac{M_3}{s-j2}$$

The coefficients are obtained using the cover-up rule as

$$M_1 = [(s+5)\,X(s)]_{s=-5}$$

$$= \left[\frac{s^2 + 10s + 4}{(s+j2)(s-j2)}\right]_{s=-5}$$

$$= \frac{-21}{29}$$

$$= -0.724$$

$$M_2 = [(s+j2)\,X(s)]_{s=-j2}$$

$$= \left[\frac{s^2 + 10s + 4}{(s+5)(s-j2)}\right]_{s=-j2}$$

$$= \frac{-j20}{-8-j20}$$

$$= 0.928\angle 21.8^\circ$$

$$M_3 = [(s-j2)\,X(s)]_{s=j2}$$

$$= \left[\frac{s^2 + 10s + 4}{(s+5)(s+j2)}\right]_{s=j2}$$

$$= \frac{j20}{-8+j20}$$

$$= 0.928\angle -21.8^\circ$$

Notice that the expansion coefficients of the complex-conjugate roots are the conjugates of each other: $M_3 = M_2^*$. This will always be the case for complex-conjugate roots. The partial fraction expansion is

$$X(s) = -\frac{0.724}{s+5} + \frac{0.928\angle 21.8^\circ}{s+j2} + \frac{0.928\angle -21.8^\circ}{s-j2}$$

You should always "sanity check" this by multiplying it out to see if any mistakes were made in the partial fraction expansion. Hence, the inverse transform is

$$x(t) = -0.724e^{-5t} + 0.928\angle 21.8^\circ e^{-j2t} + 0.928\angle -21.8^\circ e^{j2t}$$

Although this answer is correct, it is not in a form that will give insight into the response. We expect that the part containing the complex exponentials $e^{j2t}$ and $e^{-j2t}$ can be put into a sine or cosine form, so we group these to obtain

$$x(t) = -0.724e^{-5t} + 0.928\angle 21.8° e^{-j2t} + 0.928\angle -21.8° e^{j2t}$$
$$= -0.724e^{-5t} + 0.928e^{-j(2t-21.8°)} + 0.928e^{j(2t-21.8°)}$$
$$= -0.724e^{-5t} + 0.928\underbrace{e^{j(2t-21.8°)} + e^{-j(2t-21.8°)}}_{2\cos(2t-21.8°)}$$
$$= -0.724e^{-5t} + 1.857\cos(2t-21.8°)$$

which is the answer obtained for this differential equation by the direct methods of Chapter 4. As a "sanity check," we note that this evaluates to

$$x(0) = -0.724 + 1.857\cos(-21.8°)$$
$$= 1$$

and it satisfies the initial condition (as it should).

---

## Example

Solve the second-order differential equation

$$\frac{d^2x(t)}{dt^2} + 3\frac{dx(t)}{dt} + 2x(t) = 10 \qquad x(0) = 2, \quad \dot{x}(0) = 1$$

The transform of this is

$$[s^2 X(s) - 2s - 1] + 3[sX(s) - 2] + 2X(s) = \frac{10}{s}$$

Solving this for $X(s)$ gives, as was obtained in a previous example,

$$X(s) = \frac{2s^2 + 7s + 10}{s(s^2 + 3s + 2)}$$
$$= \frac{2s^2 + 7s + 10}{s(s+1)(s+2)}$$
$$= \frac{M_1}{s} + \frac{M_2}{s+1} + \frac{M_3}{s+2}$$

The expansion coefficients are obtained with the cover-up rule:

$$M_1 = [sX(s)]_{s=0}$$

$$= \left[ \frac{2s^2 + 7s + 10}{(s+1)(s+2)} \right]_{s=0}$$

$$= \frac{10}{1 \cdot 2}$$

$$= 5$$

$$M_2 = [(s+1)X(s)]_{s=-1}$$

$$= \left[ \frac{2s^2 + 7s + 10}{s(s+2)} \right]_{s=-1}$$

$$= \frac{5}{-1 \cdot 1}$$

$$= -5$$

$$M_3 = [(s+2)X(s)]_{s=-2}$$

$$= \left[ \frac{2s^2 + 7s + 10}{s(s+1)} \right]_{s=-2}$$

$$= \frac{4}{-2 \cdot -1}$$

$$= 2$$

Hence, the partial fraction expansion is

$$X(s) = \frac{5}{s} - \frac{5}{(s+1)} + \frac{2}{(s+2)}$$

You should always multiply this out to see if there are any mistakes in the partial fraction expansion. The inverse transform is

$$x(t) = 5 - 5e^{-t} + 2e^{-2t}$$

"Sanity checks" should be made to verify that this solution satisfies the initial conditions:

$$x(0) = 5 - 5 + 2 = 2$$

$$\dot{x}(0) = 5 - 4 = 1$$

which it does.

There is one final important case that needs examining: *repeated roots*. If a root is repeated, the proper partial fraction expansion is

$$F(s) = \frac{N(s)}{(s+p)^2}$$

$$= \frac{M_1}{s+p} + \frac{M_2}{(s+p)^2}$$

Observe that the correct form of the partial fraction expansion for a repeated root is that you must have all lower powers of that root represented. To obtain $M_1$ and $M_2$, multiply both sides by $(s+p)^2$ to give

$$(s+p)^2 F(s) = (s+p) M_1 + M_2$$

Evaluating this at $s = -p$ gives $M_2$:

$$\left[(s+p)^2 F(s)\right]_{s=-p} = M_2$$

Hence, for repeated roots you can obtain $M_2$ using the cover-up rule. But the expansion coefficient $M_1$ *cannot* be obtained using the cover-up rule. To obtain $M_1$, we simply evaluate the expansion for any convenient value of $s$ *except for an $s$ that is a root of the denominator:*

$$[F(s)]_{s \neq -p} = \left[ \frac{M_1}{s+p} + \frac{\overbrace{M_2}^{\text{known}}}{(s+p)^2} \right]_{s \neq -p}$$

---

## Example

Solve the following differential equation:

$$\frac{d^2 x(t)}{dt^2} + 6\frac{dx(t)}{dt} + 9x(t) = 2t \qquad x(0) = 2, \quad \dot{x}(0) = 1$$

Transforming the differential equation gives

$$[s^2 X(s) - 2s - 1] + 6[sX(s) - 2] + 9X(s) = \frac{2}{s^2}$$

Solving for $X(s)$ gives

$$X(s) = \frac{2s^3 + 13s^2 + 2}{s^2 (s+3)^2}$$

$$= \frac{M_1}{s} + \frac{M_2}{s^2} + \frac{M_3}{s+3} + \frac{M_4}{(s+3)^2}$$

Notice in this problem that two roots are repeated, $s = 0$ and $s = -3$. Hence, observe the assumed form of the partial fraction expansion. We can solve for $M_2$ and $M_4$ using the cover-up rule:

$$M_2 = [s^2 X(s)]_{s=0}$$

$$= \left[ \frac{2s^3 + 13s^2 + 2}{(s+3)^2} \right]_{s=0}$$

$$= \frac{2}{9}$$

$$M_4 = [(s+3)^2 X(s)]_{s=-3}$$

$$= \left[ \frac{2s^3 + 13s^2 + 2}{s^2} \right]_{s=-3}$$

$$= \frac{65}{9}$$

Hence, the partial fraction expansion so far is

$$X(s) = \frac{2s^3 + 13s^2 + 2}{s^2 (s+3)^2}$$

$$= \frac{M_1}{s} + \frac{2/9}{s^2} + \frac{M_3}{s+3} + \frac{65/9}{(s+3)^2}$$

We can substitute any two values of $s$ *except* $s = 0$ *and* $s = -3$. For convenience, choose $s = 1$ and $s = -1$. Substituting gives two equations to solve for $M_1$ and $M_3$:

$$\frac{17}{16} = M_1 + \frac{2}{9} + \frac{1}{4} M_3 + \frac{65}{144}$$

and

$$\frac{13}{4} = -M_1 + \frac{2}{9} + \frac{1}{2} M_3 + \frac{65}{36}$$

Rearranging these equations gives

$$4M_1 + M_3 = 1.556$$

$$-2M_1 + M_3 = 2.444$$

Solving these by the methods of Chapter 3 gives $M_1 = -0.148$ and $M_3 = 2.148$. Hence, the complete partial fraction expansion is

$$X(s) = -\frac{0.148}{s} + \frac{0.222}{s^2} + \frac{2.148}{s+3} + \frac{7.222}{(s+3)^2}$$

and the solution is

$$x(t) = -0.148 + 0.222t + 2.148e^{-3t} + 7.222te^{-3t}$$

Checking this to ensure that the initial conditions are satisfied gives

$$x(0) = -0.148 + 2.148 = 2$$

and

$$\dot{x}(0) = 0.222 - 3 \times 2.148 + 7.222 = 1$$

so the initial conditions are satisfied.

# 9 Mathematics of Vectors

Vectors are used throughout all disciplines of engineering to describe physical variables that have both a value and a direction of effect. For example, in electrical engineering these vectors represent the basic electromagnetic field vectors of the electric field intensity vector, **E**, the electric flux density vector, **D**, the magnetic field intensity vector, **H**, and the magnetic flux density vector, **B**. In mechanical engineering, vectors represent, for example, heat flow and fluid flow.

The equations governing these physical systems, are vector differential equations. We first review the familiar algebra of vectors: addition, subtraction, and products of vectors. Next we discuss the calculus of vectors: differentiation and integration. There are many possible types of differentiation and integration of vector quantities. We only discuss those found most frequently in engineering applications.

## 9.1 VECTORS AND COORDINATE SYSTEMS

A vector, as distinguished from a scalar, has two pieces of information: a value and a direction of effect. A vector is shown in the figures as a line with an arrowhead to show that direction of effect and is denoted in this chapter in boldface type (e.g., **F**). The magnitude or length of a vector will be denoted as $F$ or as $F = |\mathbf{F}|$. To compute with vectors properly requires a coordinate system. We use the rectangular (Cartesian) coordinate system, which consists of three axes, $x$, $y$, and $z$, as shown in Figure 9.1.

These axes are mutually orthogonal. In a rectangular coordinate system, a vector is described as

$$\mathbf{F} = F_x \mathbf{a}_x + F_y \mathbf{a}_y + F_z \mathbf{a}_z \tag{9.1}$$

where the components of **F** along (projections of **F** *onto*) the $x$, $y$, and $z$ axes are denoted as $F_x$, $F_y$, and $F_z$, and the unit vectors along the axes are denoted as $\mathbf{a}_x$, $\mathbf{a}_y$, and $\mathbf{a}_z$. These unit vectors are of unit length and are shown in the direction of increasing value of the coordinate axis. There are other possible coordinate systems, such as the cylindrical and spherical coordinate systems,

*Essential Math Skills for Engineers*, By Clayton R. Paul
Copyright © 2009 John Wiley & Sons, Inc.

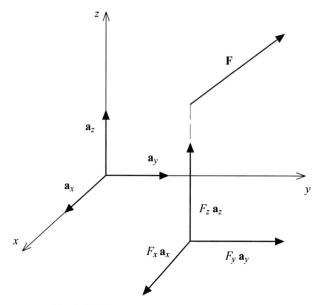

**Fig. 9.1.** The rectangular coordinate system.

but the rectangular coordinate system is the simplest and illustrates the basic ideas that are common to all other coordinate systems. Vectors are added or subtracted by adding or subtracting their corresponding components:

$$\mathbf{A} \pm \mathbf{B} = (A_x \pm B_x)\mathbf{a}_x + (A_y \pm B_y)\mathbf{a}_y + (A_z \pm B_z)\mathbf{a}_z \qquad (9.2)$$

There are two ways of performing the multiplication of two vectors: the dot product and the cross product. The *dot product* of two vectors gives the result as a *scalar* and is defined by

$$\mathbf{A} \cdot \mathbf{B} = AB\cos\theta_{AB}$$
$$= A_x B_x + A_y B_y + A_z B_z \qquad (9.3)$$

where $\theta_{AB}$ is the *smallest* angle between the two vectors, as illustrated in Figure 9.2(a). In plain terms this gives (1) the product of the length of **A** and the *projection of* **B** *onto* **A**, or (2) the product of the length of **B** and the *projection of* **A** *onto* **B**. The result for the dot product in terms of the vector components in a rectangular coordinate system given in (9.3) is easy to remember: It is the sum of the products of the corresponding components of the two vectors. Two vectors are *mutually perpendicular* if $\mathbf{A} \cdot \mathbf{B} = 0$. Also, the dot product of a vector with itself is its magnitude squared: $\mathbf{A} \cdot \mathbf{A} = |\mathbf{A}|^2$.

The *cross product* of two vectors gives a *vector* and is defined by

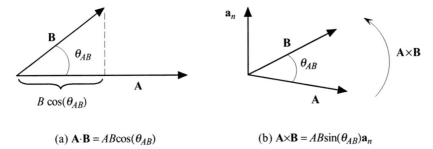

(a) $\mathbf{A \cdot B} = AB\cos(\theta_{AB})$        (b) $\mathbf{A \times B} = AB\sin(\theta_{AB})\mathbf{a}_n$

**Fig. 9.2.** The dot and cross products of two vectors.

$$\begin{aligned}
\mathbf{A \times B} &= AB\sin\theta_{AB}\,\mathbf{a}_n \\
&= (A_y B_z - A_z B_y)\mathbf{a}_x + (A_z B_x - A_x B_z)\mathbf{a}_y + (A_x B_y - A_y B_x)\mathbf{a}_z
\end{aligned} \tag{9.4}$$

where $\theta_{AB}$ is the *smallest* angle between the two vectors, as illustrated in Figure 9.2(b). The result gives a *vector* that is *perpendicular to the plane containing A and B*. The unit vector perpendicular to (normal to) this plane containing $\mathbf{A}$ and $\mathbf{B}$ is denoted as $\mathbf{a}_n$. Since there are two sides to this plane that contains $\mathbf{A}$ and $\mathbf{B}$, the direction of the unit normal is determined by the *right-hand rule*; that is, if the fingers of a person's right hand curl *from $\mathbf{A}$ to $\mathbf{B}$*, the direction of the normal to this plane for $\mathbf{A} \times \mathbf{B}$ will be given by the thumb of the right hand. The reader should practice doing this, as the concept will be used throughout the chapter. The axes of the rectangular coordinate system are assumed to be ordered cyclically according to the convention $x \to y \to z \to x \to y \to z \to \dots$ . In other words, if we cross the $x$ axis into the $y$ axis, we get the $z$ axis: $\mathbf{a}_x \times \mathbf{a}_y = \mathbf{a}_z$. Note that, for example, $\mathbf{a}_y \times \mathbf{a}_x = -\mathbf{a}_z$. The vector result for the cross product in terms of the vector components given in (9.4) for a rectangular coordinate system is easily remembered. Each component is of the form $(A_\beta B_\gamma - A_\gamma B_\beta)\mathbf{a}_\alpha$ in the order $\alpha \to \beta \to \gamma \to \alpha \to \beta \to \dots$ according to the cyclic ordering of the axes. Two vectors are *parallel* if $\mathbf{A} \times \mathbf{B} = 0$. Note that $\mathbf{A} \cdot \mathbf{B} = \mathbf{B} \cdot \mathbf{A}$ and the order in the dot product does not matter. However, the order in the cross product does matter: $\mathbf{A} \times \mathbf{B} = -\mathbf{B} \times \mathbf{A}$.

---

*Example*

Two vectors lying in the *yz* plane are defined, as shown in Figure 9.3, as

$$\mathbf{A} = 3\mathbf{a}_y$$

$$\mathbf{B} = 2\mathbf{a}_y + \mathbf{a}_z$$

The lengths of the two vectors are $A = 3$ and $B = \sqrt{(2)^2 + (1)^2} = \sqrt{5}$. The dot product is

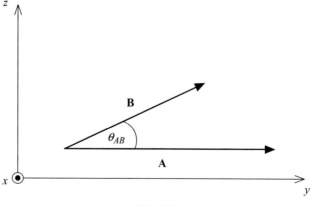

**Fig. 9.3**

$$\mathbf{A} \cdot \mathbf{B} = (0\mathbf{a}_x + 3\mathbf{a}_y + 0\mathbf{a}_z) \cdot (0\mathbf{a}_x + 2\mathbf{a}_y + \mathbf{a}_z)$$
$$= 3 \times 2 + 0 \times 1$$
$$= 6$$

From the dot product in (9.3),

$$\cos\theta_{AB} = \frac{\mathbf{A} \cdot \mathbf{B}}{AB}$$
$$= \frac{6}{3 \cdot \sqrt{5}}$$
$$= 0.894$$

Hence, the smallest angle between the two vectors is $\theta_{AB} = \cos^{-1}(0.894) = 26.57°$. For these simple vectors, we can obtain this angle directly by trigonometry:

$$\theta_{AB} = \tan^{-1}\frac{1}{2}$$
$$= 26.57°$$

so again, we obtain

$$\mathbf{A} \cdot \mathbf{B} = AB\cos\theta_{AB}$$
$$= 3\sqrt{2^2 + 1^2}\,\cos(26.57°)$$
$$= 6$$

The cross product is

$$\mathbf{A} \times \mathbf{B} = (A_y B_z - A_z B_y)\mathbf{a}_x + (A_z B_x - A_x B_z)\mathbf{a}_y + (A_x B_y - A_y B_x)\mathbf{a}_z$$
$$= (3 - 0)\mathbf{a}_x + (0 - 0)\mathbf{a}_y + (0 - 0)\mathbf{a}_z$$
$$= 3\mathbf{a}_x$$

Directly, we obtain the same result:

$$\mathbf{A} \times \mathbf{B} = AB\sin\theta_{AB}\,\mathbf{a}_n$$
$$= 3\sqrt{2^2 + 1^2}\,\sin(26.57°)\mathbf{a}_n$$
$$= 3\mathbf{a}_n$$

Since both vectors lie in the *yz* plane, the unit normal *perpendicular to the plane containing* **A** and **B** is in the ±*x* direction. Using the right-hand rule and crossing **A** *to* **B** gives the unit normal in the positive *x* direction: $\mathbf{a}_n = \mathbf{a}_x$.

---

## 9.2  THE LINE INTEGRAL

The equations in engineering systems that involve vectors generally involve two basic integrals: the *line integral* and the *surface integral*. Hence, it is important that we understand what these mean and how to evaluate them. The vectors in the physical equations are functions of the coordinate system variables *x*, *y*, and *z*, denoted $\mathbf{F}(x, y, z)$, and hence are said to constitute a *field*. An example of a *scalar field* is a plot of the temperature distribution in a room. Contours of constant temperature (a scalar) show the distribution of that field in the room. An example of a *vector field* would be the plot of flow rates and directions of the water flow in a river. The directions of these vectors show the direction of the water flow at that point, and the lengths of these vectors are proportional to the rates of flow at that point.

The *line integral* of a vector field is denoted as

$$\boxed{\int_a^b \mathbf{F}(x, y, z)\cdot d\mathbf{l} = \int_a^b F(x, y, z)\cos\theta\,dl} \tag{9.5}$$

The line integral means that we take the products of the projection of the vector **F** onto the path, $F\cos\theta$ (alternatively, the component of **F** tangent to the path) and the differential lengths, *dl*, along the path, and sum them with an integral from the starting point *a* to the endpoint *b*, as illustrated in Figure 9.4. An example of a line integral is the computation of the work required to push an object from one point to another when the force, **F**, is exerted on the object at an angle to the path as shown in Figure 9.5. The work done is $W = \int|F|\cos\theta\,dx = \int\mathbf{F}\cdot d\mathbf{l}$. The line integral is a very sensible result. There are two components of **F**: one component *parallel* to the path and the other component *perpendicular* to the path. Only the component *parallel to the path* should contribute to the result.

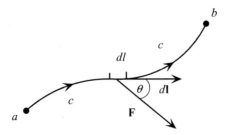

**Fig. 9.4.** The line integral.

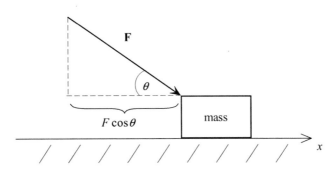

**Fig. 9.5.** The line integral in computing work.

The actual computation of the line integral in a rectangular coordinate system is very simple. In a rectangular coordinate system a vector differential path length is

$$dl = dx\,\mathbf{a}_x + dy\,\mathbf{a}_y + dz\,\mathbf{a}_z \tag{9.6}$$

Hence,

$$\mathbf{F}\cdot d\mathbf{l} = F_x\,dx + F_y\,dy + F_z\,dz \tag{9.7}$$

and the line integral becomes

$$\boxed{\begin{aligned}\int_a^b \mathbf{F}\cdot d\mathbf{l} &= \int_a^b F\cos\theta\,dl \\ &= \int_{x_a}^{x_b} F_x\,dx + \int_{y_a}^{y_b} F_y\,dy + \int_{z_a}^{z_b} F_z\,dz\end{aligned}} \tag{9.8}$$

where the path extends from $(x_a, y_a, z_a)$ to $(x_b, y_b, z_b)$ and each component of $\mathbf{F}$ is a function of $x$, $y$, and $z$: $F_x(x, y, z)$, $F_y(x, y, z)$, and $F_z(x, y, z)$. If the integral is taken around a closed path, it is denoted with a circle on the integral sign as $\oint_c \mathbf{F}\cdot d\mathbf{l}$, and $c$ represents the contour of that closed path.

*Example*

A vector field in the *yz* plane is given as

$$\mathbf{F}(x, y, z) = z\mathbf{a}_y$$

as shown in Figure 9.6. Determine the line integral of $\mathbf{F}$ along a straight-line path between the two points in the *yz* plane *from* point *a* at (0,1,3) *to* point *b* at (0,2,4).

Observe that at all points in the *yz* plane the vector is directed in the *y* direction. However, its magnitude depends on *z*: For positive, increasing values of *z*, its magnitude (length) increases. For *z* negative, it is pointing in the −*y* direction. Performing the line integral gives

$$\int_a^b \mathbf{F}\cdot d\mathbf{l} = \int_{x=0}^{0}\underbrace{F_x}_{0}dx + \int_{y=1}^{2}\underbrace{F_y}_{z}dy + \int_{z=3}^{4}\underbrace{F_z}_{0}dz$$

$$= \int_{y=1}^{2} z\,dy$$

$$= \int_{y=1}^{2}(y+2)\,dy$$

$$= \frac{7}{2}$$

and we have substituted the equation of the path, *z* = *y* + 2.

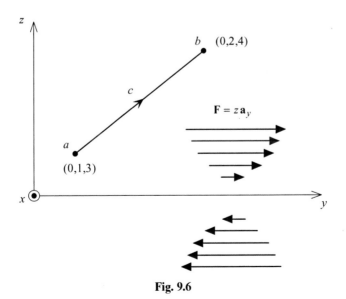

**Fig. 9.6**

## 9.3   THE SURFACE INTEGRAL

The *surface integral* is

$$\int_s \mathbf{F}(x, y, z) \cdot d\mathbf{s} = \int_s \mathbf{F}(x, y, z) \cdot \mathbf{a}_n \, ds$$
$$= \int_s F(x, y, z) \cos\theta \, ds \qquad (9.9)$$

The surface integral gives the integral of the products of the components of **F** that are *perpendicular to the surface s* and the differential surface elements *ds* as shown in Fig. 9.7. The unit normal perpendicular to the surface is denoted as $\mathbf{a}_n$, and the differential surface area is $d\mathbf{s} = ds\,\mathbf{a}_n$. The surface integral gives the *flux of the vector field* **F** *through the surface s*. This is like shining a light through an opening. There are two components of **F**: One component is *parallel* to the surface and the other component is *perpendicular* to the surface. Only the component of the light flux that is *perpendicular* to the opening contributes to the net light flux passing through that opening. If the surface *s* is a closed surface, the surface integral is denoted with a circle on the integral sign: $\oint_s \mathbf{F} \cdot d\mathbf{s}$.

Observe that there is a major difference between the line integral and the surface integral. The line integral involves the component of **F** that is *parallel* to (tangent to) the path, whereas the surface integral involves the component of **F** that is *perpendicular* to the surface.

The evaluation of the surface integral in a rectangular coordinate system is very simple. The vector differential surface is

$$d\mathbf{s} = dy\,dz\,\mathbf{a}_x + dx\,dz\,\mathbf{a}_y + dx\,dy\,\mathbf{a}_z \qquad (9.10)$$

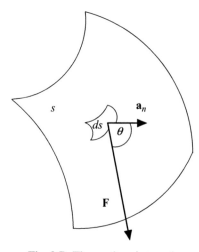

**Fig. 9.7.** The surface integral.

Note that the components of this are the differential surface areas whose unit normals are perpendicular to them (e.g., $dy\,dz\mathbf{a}_x$). Hence, the surface integral simplifies, in a rectangular coordinate system, to

$$\boxed{\int_s \mathbf{F}(x, y, z)\cdot d\mathbf{s} = \int\int F_x\,dy\,dz + \int\int F_y\,dx\,dz + \int\int F_z\,dx\,dy}\qquad (9.11)$$

---

### Example

A wedge-shaped surface lies in the $yz$ plane as shown in Figure 9.8.
Determine the flux of the vector field

$$\mathbf{F} = (x+2)\mathbf{a}_x$$

through the surface.
The surface integral becomes

$$\int_s \mathbf{F}(x, y, z)\cdot d\mathbf{s} = \int\int F_x\,dy\,dz$$
$$= \int_{z=1}^3 \int_{y=1}^{y=-z+4}(x+2)\,dy\,dz$$
$$= \int_{z=1}^3 \int_{y=1}^{y=-z+4} 2\,dy\,dz$$
$$= \int_{z=1}^3 (-2z+6)\,dz$$
$$= 4$$

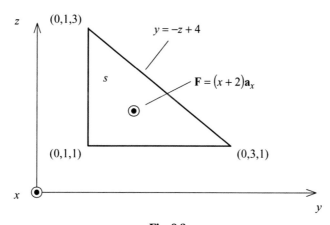

**Fig. 9.8**

We have substituted $x = 0$ over the surface into $F_x = x + 2$ and the equation of the top part of the wedge, $y = -z + 4$, in the limit of one of the integrals.

---

## 9.4  DIVERGENCE

The line and surface integrals apply over regions of space. The following vector calculus results, the *divergence* and the *curl*, are the *point forms* of these integrals which apply to points in space and are differential relations that give the relation of the field vectors at points in space.

The *divergence* of a vector field gives the net *outflow* of a vector field from a point, hence the name *divergence*, and is defined by

$$\nabla \cdot \mathbf{F}(x, y, z) = \lim_{\Delta v \to 0} \frac{\oint_s \mathbf{F} \cdot d\mathbf{s}}{\Delta v} \tag{9.12}$$

This is illustrated in Figure 9.9.

If we surround a point by a closed surface $s$ that contains a differential volume $\Delta v$, compute the net flux of $\mathbf{F}$ *out of* the closed surface per unit of volume enclosed by $s$, and then let the surface and enclosed volume shrink to zero, the limit of that is the *divergence* of $\mathbf{F}$ at that point. Essentially, this gives an indication of any sources of $\mathbf{F}$ that are located at the point. If the divergence of $\mathbf{F}$ is negative at the point, we say that a *sink* exists at that point. So the divergence indicates whether there is a net *outflow* of $\mathbf{F}$ at that point. If we puncture an inflated balloon, we get a divergence of the air contained in that ballon.

The "del operator" is somewhat equivalent to a derivative in scalar calculus and is an "operator" defined by

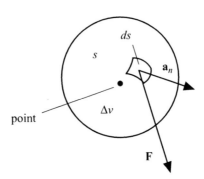

**Fig. 9.9.** Divergence of a vector field.

$$\nabla = \mathbf{a}_x \frac{\partial}{\partial x} + \mathbf{a}_y \frac{\partial}{\partial y} + \mathbf{a}_z \frac{\partial}{\partial z} \tag{9.13}$$

Using the del operator, we obtain the divergence of a vector field in a rectangular coordinate system as

$$\nabla \cdot \mathbf{F}(x, y, z) = \frac{\partial F_x}{\partial x} + \frac{\partial F_y}{\partial y} + \frac{\partial F_z}{\partial z} \tag{9.14}$$

It is very important to observe that the divergence of a vector field gives a *scalar* quantity as the result.

---

### Example

A vector field is described by

$$\mathbf{F} = x\mathbf{a}_x + y\mathbf{a}_y + z\mathbf{a}_z$$

as plotted in Figure 9.10. Determine the divergence of the field.

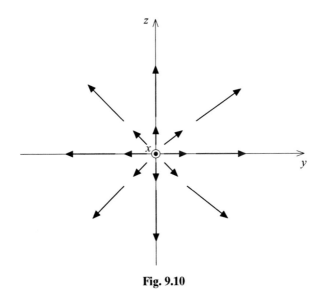

**Fig. 9.10**

The divergence of this field is

$$\nabla \cdot \mathbf{F}(x, y, z) = \frac{\partial F_x}{\partial x} + \frac{\partial F_y}{\partial y} + \frac{\partial F_z}{\partial z}$$
$$= 1 + 1 + 1 = 3$$

Since this result is independent of $x$, $y$, and $z$, there is a net outflow of the vector at every point in the space. This is a sensible result since the field is constant over any sphere of radius $r = \sqrt{x^2 + y^2 + z^2}$ centered at the origin of the coordinate system and is directed normal to the surface of that sphere. Hence, from the basic definition of the divergence given in (9.12), we can calculate directly

$$\nabla \cdot \mathbf{F}(x, y, z) = \lim_{\Delta v \to 0} \frac{\oint_s \mathbf{F} \cdot d\mathbf{s}}{\Delta v}$$
$$= \frac{r \times 4\pi r^2}{(4/3)\pi r^3}$$
$$= 3$$

---

### 9.4.1   The Divergence Theorem

We can interchange certain surface and volume integrals with the *divergence theorem*:

$$\boxed{\oint_s \mathbf{F} \cdot d\mathbf{s} = \int_v (\nabla \cdot \mathbf{F}) \, dv} \tag{9.15}$$

This result provides that if we integrate the divergence of $\mathbf{F}$ throughout some volume $v$, we can obtain the same result by performing the surface integral of $\mathbf{F}$ over the *closed* surface $s$ that contains the volume $v$. This is a very sensible result if we think about what these quantities mean. According to (9.12), the divergence $\nabla \cdot \mathbf{F}$ gives the net outflow or flux of $\mathbf{F}$ throughout the volume $\Delta v$ per unit of that volume. Rewriting (9.12) gives

$$\oint_s \mathbf{F} \cdot d\mathbf{s} = \lim_{\Delta v \to 0} [\nabla \cdot \mathbf{F}(x, y, z) \Delta v] \tag{9.12}$$
$$= \int_v (\nabla \cdot \mathbf{F}) \, dv$$

Hence, it makes sense that we can obtain the net flux out of the closed surface $s$ that encloses that volume, $\oint_s \mathbf{F} \cdot d\mathbf{s}$, by performing the volume integral of $\nabla \cdot \mathbf{F}$ throughout that volume.

## *Example*

Verify the divergence theorem for the vector field

$$\mathbf{F} = x\,\mathbf{a}_x + y\,\mathbf{a}_y + z\,\mathbf{a}_z$$

for the square volume whose corners are at $(0,0,0)$, $(0,0,1)$, $(0,1,0)$, $(0,1,1)$, $(1,0,0)$, $(1,0,1)$, $(1,1,0)$, and $(1,1,1)$ as illustrated in Figure 9.11.

The surface integral over the closed surface $s$ is

$$\oint_s \mathbf{F} \cdot d\mathbf{s} = \underbrace{\int_{z=0}^{1}\int_{y=0}^{1} F_x\, dy\, dz}_{\text{front}} - \underbrace{\int_{z=0}^{1}\int_{y=0}^{1} F_x\, dy\, dz}_{\text{back}} - \underbrace{\int_{z=0}^{1}\int_{x=0}^{1} F_y\, dx\, dz}_{\text{left}} + \underbrace{\int_{z=0}^{1}\int_{x=0}^{1} F_y\, dx\, dz}_{\text{right}} -$$

$$\underbrace{\int_{y=0}^{1}\int_{x=0}^{1} F_z\, dx\, dy}_{\text{bottom}} + \underbrace{\int_{y=0}^{1}\int_{x=0}^{1} F_z\, dx\, dy}_{\text{top}}$$

$$= \underbrace{\int_{z=0}^{1}\int_{y=0}^{1} \underset{1}{x}\, dy\, dz}_{\text{front}} - \underbrace{\int_{z=0}^{1}\int_{y=0}^{1} \underset{0}{x}\, dy\, dz}_{\text{back}} - \underbrace{\int_{z=0}^{1}\int_{x=0}^{1} \underset{0}{y}\, dx\, dz}_{\text{left}} + \underbrace{\int_{z=0}^{1}\int_{x=0}^{1} \underset{1}{y}\, dx\, dz}_{\text{right}} -$$

$$\underbrace{\int_{y=0}^{1}\int_{x=0}^{1} \underset{0}{z}\, dx\, dy}_{\text{bottom}} + \underbrace{\int_{y=0}^{1}\int_{x=0}^{1} \underset{1}{z}\, dx\, dy}_{\text{top}}$$

$$= 1 - 0 - 0 + 1 - 0 + 1$$

$$= 3$$

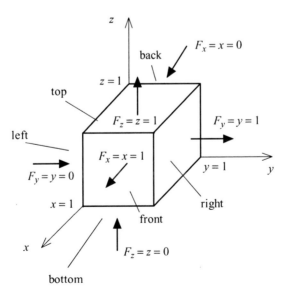

**Fig. 9.11**

Notice that the surface integral determines the *net flux leaving the closed surface*. A vector component points into one side and out of the other side. Hence, half the integrals are positive and half the integrals are negative. Observe also that each integrand is 0 or 1 over a surface and the dimensions of each side are 1. Therefore, the integral over a side is either 0 or 1. Since

$$\nabla \cdot \mathbf{F}(x, y, z) = \frac{\partial F_x}{\partial x} + \frac{\partial F_y}{\partial y} + \frac{\partial F_z}{\partial z}$$
$$= 1 + 1 + 1 = 3$$

the right-hand side of the divergence theorem in (9.15) gives the same result:

$$\int_v (\nabla \cdot \mathbf{F}) \, dv = \int_{x=0}^1 \int_{y=0}^1 \int_{z=0}^1 3 \, dx \, dy \, dz$$
$$= 3$$

---

## 9.5 CURL

While the divergence gives the net outflow or flux of a vector field from a point, the *curl* of a vector field gives the *net circulation or rotation of the field about a point*. For example, consider the vector field shown in Figure 9.12. This

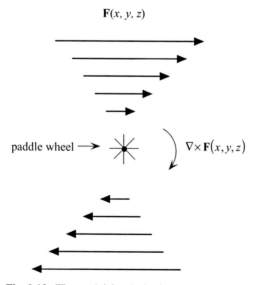

**Fig. 9.12.** The curl (circulation) of a vector field.

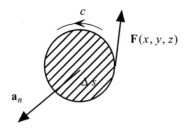

**Fig. 9.13.** Defining the curl of a vector field.

field might represent the flow of the water in a river. If we insert a small paddle wheel as shown, the flow pattern will cause the paddle wheel to rotate in the clockwise direction. If we turned the paddle wheel such that its axis was parallel to the field lines, it would not rotate.

Figure 9.13 shows how we might define the *circulation* of a vector field in one plane. Define a flat surface $s$ in that plane and the associated contour $c$ enclosing it. Define the unit normal to that plane as $\mathbf{a}_n$, with its direction according to the right-hand rule with respect to the direction of $c$ around that surface perimeter. The net circulation at the point per unit of the enclosed surface area *in this plane* would be

$$circulation\ per\ unit\ area = \mathbf{a}_n \left( \lim_{\Delta s \to 0} \frac{\oint_c \mathbf{F} \cdot d\mathbf{l}}{\Delta s} \right) \tag{9.16}$$

By performing the line integral of $\mathbf{F}$ around the contour $c$ enclosing the surface $\Delta s$ and dividing by that surface, we get a measure of the circulation (in this case in the counterclockwise direction). A direction is given to that circulation by the unit vector $\mathbf{a}_n$ normal to the surface. The direction of the unit normal is obtained in accordance with the right-hand rule. Since the result is circulation or rotation of the field, we should obtain the total circulation or rotation in three orthogonal planes. The result gives the curl of the vector field as

$$\nabla \times \mathbf{F}(x, y, z) = \mathbf{a}_x \left( \lim_{\Delta s_{yz} \to 0} \frac{\oint_{c_{yz}} \mathbf{F} \cdot d\mathbf{l}}{\Delta s_{yz}} \right) + \mathbf{a}_y \left( \lim_{\Delta s_{xz} \to 0} \frac{\oint_{c_{xz}} \mathbf{F} \cdot d\mathbf{l}}{\Delta s_{xz}} \right) + \mathbf{a}_z \left( \lim_{\Delta s_{xy} \to 0} \frac{\oint_{c_{xy}} \mathbf{F} \cdot d\mathbf{l}}{\Delta s_{xy}} \right)$$

$$\tag{9.17}$$

where $\Delta s_{xy}$, for example, is a flat surface in the $xy$ plane which is perpendicular to $\mathbf{a}_z$, and $c_{xy}$ denotes the contour around the enclosed surface $\Delta s_{xy}$.

Applying the del operator that is defined in (9.13) gives a mechanical way of determining the curl in a rectangular coordinate system:

$$\nabla \times \mathbf{F}(x, y, z) = \left( \frac{\partial F_z}{\partial y} - \frac{\partial F_y}{\partial z} \right) \mathbf{a}_x + \left( \frac{\partial F_x}{\partial z} - \frac{\partial F_z}{\partial x} \right) \mathbf{a}_y + \left( \frac{\partial F_y}{\partial x} - \frac{\partial F_x}{\partial y} \right) \mathbf{a}_z \qquad (9.18)$$

Observe that each of these components can easily be remembered using the cyclic rule for the cross product, the cyclic ordering of the three axes, and the definition of the del operator given in (9.13). For example, each component of the curl is of the form $\left( \frac{\partial F_\gamma}{\partial \beta} - \frac{\partial F_\beta}{\partial \gamma} \right) \mathbf{a}_\alpha$, where the ordering is $\alpha \to \beta \to \gamma \to \alpha \dots$ .

---

### Example

Determine the curl of the vector field

$$\mathbf{F} = z \, \mathbf{a}_y$$

which is illustrated in Figure 9.14.

First, we see clearly that there will be circulation and the rotation will be clockwise with the unit normal being in the negative $x$ direction. Substituting into (9.18) yields

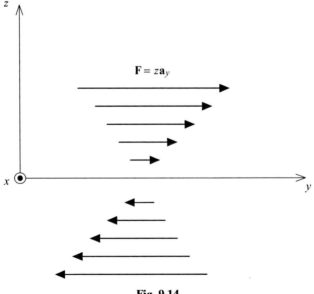

**Fig. 9.14**

$$\nabla \times \mathbf{F}(x, y, z) = \underbrace{\left( \frac{\partial F_z}{\partial y} - \frac{\partial F_y}{\partial z} \right)}_{0}\mathbf{a}_x + \left( \underbrace{\frac{\partial F_x}{\partial z}}_{1} - \underbrace{\frac{\partial F_z}{\partial x}}_{0} \right)\mathbf{a}_y + \left( \underbrace{\frac{\partial F_y}{\partial x}}_{0} - \underbrace{\frac{\partial F_x}{\partial y}}_{0} \right)\mathbf{a}_z$$

$$= -\mathbf{a}_x$$

as expected.

---

### 9.5.1  Stokes' Theorem

Similar to the divergence theorem, *Stokes' theorem* allows us to interchange a surface integral and a line integral:

$$\boxed{\oint_c \mathbf{F} \cdot d\mathbf{l} = \int_s (\nabla \times \mathbf{F}) \cdot d\mathbf{s}} \qquad (9.19)$$

Stokes' theorem provides that the surface integral of the curl of **F** over an open surface *s* will give the same result as performing the line integral of **F** around the contour *c* that encloses that open surface. As was the case for the divergence theorem, Stokes' theorem is a very sensible result. According to (9.17), the curl of a vector field, $\nabla \times \mathbf{F}$, gives the net circulation or rotation of a field around a contour that encloses a differential surface per unit of that enclosed surface. Rewriting the *x* component of (9.17) gives

$$\oint_{c_{yz}} \mathbf{F} \cdot d\mathbf{l} = \lim_{\Delta s_{yz} \to 0} \{ [\nabla \times \mathbf{F}(x, y, z)]_x \, \Delta s_{yx} \}$$

$$= \int_{s_{yz}} (\nabla \times \mathbf{F}) \cdot d\mathbf{s}$$

Hence, it makes sense that by integrating the curl over the surface with a surface integral, we will obtain the same result as the line integral around the contour enclosing that surface would give.

---

### *Example*

Verify Stokes' theorem for the vector field

$$\mathbf{F} = z\,\mathbf{a}_y$$

and the closed contour *c* and its enclosed surface *s* shown in Figure 9.15. The curl of **F** is

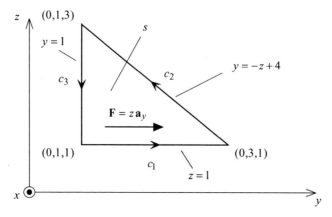

**Fig. 9.15**

$$\nabla \times \mathbf{F} = \left( \underbrace{\frac{\partial F_z}{\partial y}}_{0} - \underbrace{\frac{\partial F_y}{\partial z}}_{1} \right) \mathbf{a}_x + \left( \underbrace{\frac{\partial F_x}{\partial z}}_{0} - \underbrace{\frac{\partial F_z}{\partial x}}_{0} \right) \mathbf{a}_y + \left( \underbrace{\frac{\partial F_y}{\partial x}}_{0} - \underbrace{\frac{\partial F_x}{\partial y}}_{0} \right) \mathbf{a}_z$$

$$= -\mathbf{a}_x$$

Hence, the right-hand side of Stokes' theorem is

$$\int_s (\nabla \times \mathbf{F}) \cdot d\mathbf{s} = \int_{z=1}^{3} \int_{y=1}^{y=-z+4} \underbrace{(-1)\mathbf{a}_x}_{\nabla \times \mathbf{F}} \cdot \underbrace{(\mathbf{a}_x \, dy \, dz)}_{d\mathbf{s}}$$

$$= \int_{z=1}^{3} \int_{y=1}^{y=-z+4} (-1) \, dy \, dz$$

$$= -2$$

Since $\mathbf{F} \cdot d\mathbf{l} = F_y \, dy = z \, dy$, the left-hand side of Stokes' theorem is

$$\oint_c \mathbf{F} \cdot d\mathbf{l} = \underbrace{\int_{y=1}^{3} F_y \, dy}_{c_1} + \underbrace{\int_{y=3}^{1} F_y \, dy}_{c_2} + \underbrace{\int_{y=1}^{1} F_y \, dy}_{c_3}$$

$$= \int_{y=1}^{3} \underbrace{z}_{1} \, dy + \int_{y=3}^{1} \underbrace{z}_{-y+4} \, dy + \int_{y=1}^{1} z \, dy$$

$$= 2 - 4 + 0$$

$$= -2$$

which is the same.

## 9.6  THE GRADIENT OF A SCALAR FIELD

Perhaps one of the best illustrations of the use of the *gradient* is a topographical map. Contours of constant elevation (above sea level) are shown as closed contours. We might denote this as the scalar field $EL(x, y, z)$. Think of this scalar function as depicting a three-dimensional map with the $x$ and $y$ coordinates giving the horizontal position over the Earth's surface, and the $z$ axis giving the elevation of each point above sea level. The closer the contours of constant elevation are to each other, the steeper the slope (i.e., the greater the change in elevation with a change in horizontal distance). If we wanted to chart a course for hiking that would avoid the steep slopes, we would choose a path between points on adjacent contours of constant elevation with those contours being as widely separated as possible. In doing so, we would make the vertical distance we move as long a horizontal distance as possible. Also, to make the trip as expeditious as possible, we would choose a route that is perpendicular to those contours.

Denote some general scalar field as $f(x, y, z)$. A differential change in the function (the scalar field) as we move between contours of constant value of $f$ is

$$df = \frac{\partial f(x, y, z)}{\partial x}dx + \frac{\partial f(x, y, z)}{\partial y}dy + \frac{\partial f(x, y, z)}{\partial z}dz \qquad (9.20)$$

Using the del operator given in (9.13),

$$\nabla = \mathbf{a}_x \frac{\partial}{\partial x} + \mathbf{a}_y \frac{\partial}{\partial y} + \mathbf{a}_z \frac{\partial}{\partial z} \qquad (9.13)$$

we define the *gradient* of $f$ as

$$\boxed{\nabla f = \frac{\partial f(x, y, z)}{\partial x}\mathbf{a}_x + \frac{\partial f(x, y, z)}{\partial y}\mathbf{a}_y + \frac{\partial f(x, y, z)}{\partial z}\mathbf{a}_z} \qquad (9.21)$$

Note that the gradient of a scalar field $f(x, y, z)$, $\nabla f(x, y, z)$ gives a *vector* as the result. Recalling the vector differential path length given in (9.6),

$$d\mathbf{l} = dx\,\mathbf{a}_x + dy\,\mathbf{a}_y + dz\,\mathbf{a}_z \qquad (9.6)$$

we can write (9.20) in terms of the gradient as

$$\boxed{df = \nabla f \cdot d\mathbf{l}} \qquad (9.22)$$

which you should verify.

Next, we interpret the meaning of the gradient. The differential change in (9.22) is

$$df = \nabla f \cdot d\mathbf{l} \tag{9.23}$$
$$= |\nabla f| dl \cos \theta$$

where $\theta$ is the angle between the gradient vector, $\nabla f$, and the differential path length vector, $d\mathbf{l}$. The rate of change of the scalar field along this path is

$$\frac{df}{dl} = |\nabla f| \cos \theta \tag{9.24}$$

If we want to move in the direction of the maximum rate of change of the scalar field (i.e., perpendicular to the contours of constant $f$), the path taken must be perpendicular to the gradient vector (i.e., $\theta = 90°$):

$$\boxed{\left. \frac{df}{dl} \right|_{max} = |\nabla f|} \tag{9.25}$$

Therefore, *the gradient vector gives both the direction and the magnitude of the maximum space rate of change of the scalar field.*

---

### Example

Show that the gradient of the scalar field $f(x, y, z) = x + y$ is normal to the lines of constant $f$.

The scalar field is plotted in Fig. 9.16. The gradient is

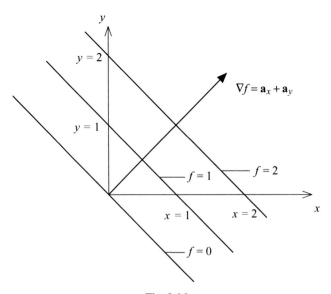

**Fig. 9.16**

$$\nabla f = \frac{\partial f}{\partial x}\mathbf{a}_x + \frac{\partial f}{\partial y}\mathbf{a}_y$$

$$= \mathbf{a}_x + \mathbf{a}_y$$

which is plotted in Figure 9.16. Obviously, the gradient is perpendicular to the lines of constant $f$, and it also points in the direction of the maximum rate of change of $f$.

# INDEX

*Essential Math Skills for Engineers*, By Clayton R. Paul
Copyright © 2009 John Wiley & Sons, Inc.

Breinigsville, PA USA
21 August 2010
244001BV00002B/2/P